A Recipe for Success Using SAS® University Edition

How to Plan Your First Analytics Project

Sharon Torrence Jones

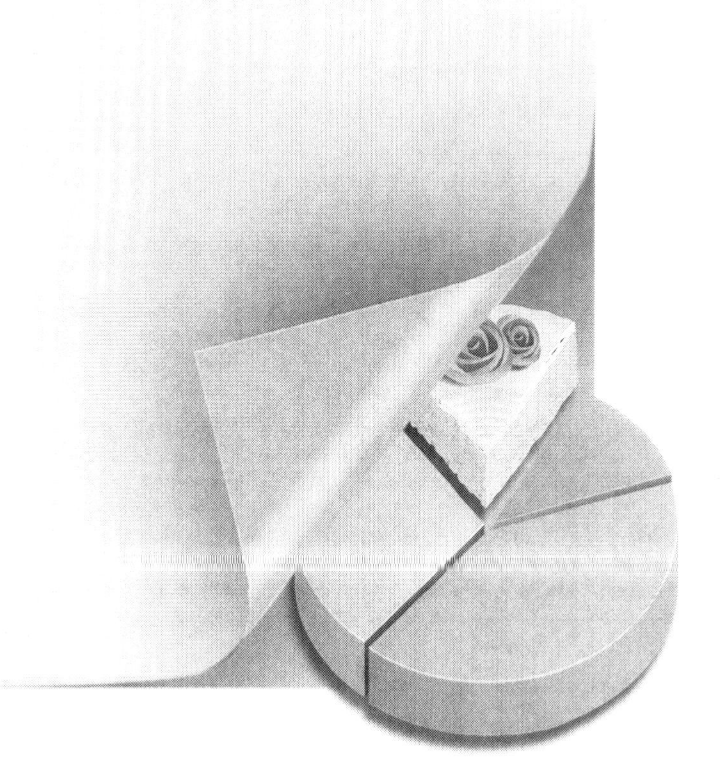

support.sas.com/bookstore

The correct bibliographic citation for this manual is as follows: Jones, Sharon Torrence. 2016. *A Recipe for Success Using SAS® University Edition: How to Plan Your First Analytics Project*. Cary, NC: SAS Institute Inc.

A Recipe for Success Using SAS® University Edition: How to Plan Your First Analytics Project

ISBN 978-1-62960-193-9 (Hard copy)
ISBN 978-1-62960-194-6 (Epub)
ISBN 978-1-62960-195-3 (Mobi)
ISBN 978-1-62960-196-0 (PDF)

SAS Institute Inc., SAS Campus Drive, Cary, NC 27513-2414

August 2016

Contents

About This Book

Purpose

This book is for the person that has an interest in analytics and knows about SAS®, but is not exactly sure how to get started, or, for the person who has an interest in data and wants to know more!

The book will take you on a journey through how to start using analytics, how to use SAS to accomplish a goal, how to apply SAS to your community and where it is effective, and provide real case studies of how users like you implemented SAS to solve their problems.

Is This Book for You?

This book is for anyone interested in understanding how to use SAS for real world applications. This book is great for students, in the classroom, and for the average person who wants to learn to use SAS, but has little to no experience in planning, creating, or executing a programming or analytics project.

Scope of This Book

The book will give users the ability to use SAS programs for their betterment. With data analysis at the forefront of business and more emphasis being placed on data-driven decisions, this book will help illustrate how you can use data to solve real world problems. This book will introduce you to using SAS and the power the software brings to data analytics and will provide a guide on using analytics. The book does not go into extensive detail of statistical concepts, but instead introduces them and provides an overview in the context of using SAS.

About the Examples

Software Used to Develop the Book's Content

If you are using SAS University Edition to access data and run your programs, then please check the SAS University Edition page to ensure that the software contains the product or products that you need to run the code: **http://support.sas.com/software/products/university-edition/index.html**

Example Code and Data

You can access the example code and data for this book by linking to its author page at **http://support.sas.com/publishing/authors**. Select the name of the author. Then, look for the cover thumbnail of this book, and select Example Code and Data to display the SAS programs that are included in this book.

If you are unable to access the code through the Web site, send e-mail to saspress@sas.com.

Additional Resources

Data is everywhere and has become such a large part of our lives. It can seem overwhelming, but SAS University Edition is a great platform to start exploring how data and analytics can change how you solve problems and make decisions. I was a programming novice when I began to learn SAS, and I was nervous about being able to completely understand analytics. But once I got started, I found that the syntax and logic were easy to learn and implement. I was able to create new knowledge by manipulating data that I was using in my everyday life! My hope is that as you move through this book, you will learn how to use SAS and apply to a real world project. We all have the power to transform data and create knowledge, and SAS University Edition can help you begin your journey.

To begin to explore a path in data science, there are educational resources for all ages and skill levels available for free on the web.

- Code.org **https://code.org/**
- MIT's Scratch project **https://scratch.mit.edu/**

- Code Academy **https://www.codeacademy.com/**
- Khan Academy's computer programming courses **https://www.khanacademy.org/computing/computer-programming**
- A SAS high school programming course **http://support.sas.com/learn/ap/hs/index.html**
- SAS University Edition **http://www.sas.com/en_us/software/university-edition.html**
- Online programming tutorials **https://support.sas.com/edu/schedules.html?ctry=us=2588**
- Online statistics tutorials **https://support.sas.com/edu/schedules.html?id=1320=US**

Use your passion as fuel and find the data science path that fits you!

Although this book illustrates many analyses regularly performed in businesses across industries, questions specific to your aims and issues may arise. To fully support you, SAS Institute and SAS Press offer you the following help resources:

- For questions about topics covered in this book, contact the author through SAS Press:
  - Send questions by email to saspress@sas.com; include the book title in your correspondence.
  - Submit feedback on the author's page at **http://support.sas.com/author_feedback**.
- For questions about topics in or beyond the scope of this book, post queries to the relevant SAS Support Communities at **https://communities.sas.com/welcome**.
- SAS Institute maintains a comprehensive website with up-to-date information. One page that is particularly useful to both the novice and the seasoned SAS user is its Knowledge Base. Search for relevant notes in the "Samples and SAS Notes" section of the Knowledge Base at **http://support.sas.com/resources**.
- • Registered SAS users or their organizations can access SAS Customer Support at **http://support.sas.com**. Here you can pose specific questions to SAS Customer Support; under Support, click Submit a Problem. You will need to provide an email address to which replies can be sent, identify your organization, and provide a customer site number or license information. This information can be found in your SAS logs.

Keep in Touch

We look forward to hearing from you. We invite questions, comments, and concerns. If you want to contact us about a specific book, please include the book title in your correspondence.

To Contact the Author through SAS Press

By e-mail: saspress@sas.com

Via the Web: http://support.sas.com/author_feedback

SAS Books

For a complete list of books available through SAS, visit http://support.sas.com/bookstore.

Phone: 1-800-727-3228

Fax: 1-919-677-8166

E-mail: sasbook@sas.com

SAS Book Report

Receive up-to-date information about all new SAS publications via e-mail by subscribing to the SAS Book Report monthly eNewsletter. Visit http://support.sas.com/sbr.

Publish with SAS

SAS is recruiting authors! Are you interested in writing a book? Visit http://support.sas.com/saspress for more information.

About the Author

Sharon Jones Sharon Jones, Ed.D, is a faculty member at Central Piedmont Community college in Charlotte, NC, where she leads and continues to develop the SAS curriculum in the Continuing and Corporate Education department. Dr. Jones has been in education for 13 years as a Career and Technical Education teacher in the Charlotte Mecklenburg Schools, the Wake County Schools, and as an industry trainer. She has taught courses in computer programming, web design, ecommerce, computer science principles, and SAS programming. She has also presented and been published nationally and internationally on SAS and on integrating technology tools in the classroom.

Acknowledgments

This book has been a personal journey and a challenge for me. I was not sure how the book would come together, and the process of putting my thoughts down was much more involved than I could ever imagine, but it has been a wonderful experience. I wrote this book because I truly believe in SAS and the Power to Know.

I want to thank my family, my husband Ricky, and my children for their support while I worked on this book. I also want to thank my parents for their endless love and support and to my colleagues who continue to encourage me every day. I would also like to thank the SAS Press team: my copyeditor Amy Wolfe, graphic designer Robert Harris, and production specialist Denise Jones. A huge thanks to my SAS Press developmental editor, Brenna Leath, for her vision and faith in me. I thank you for this amazing opportunity!

1

The SAS Language

Structure of the Book

Do you hear the phrase, The Power of SAS®, but you really don't know what it means and what it could mean for you? Do you wonder how you can use SAS to make your life or job easier, but you do not know where to start? Well, you have come to the right place! This book is an easy guide on how to start using SAS and apply it to the real world. It is filled with helpful examples and real-life success stories of SAS users.

This book shows you:

- how to start using analytics
- how to use SAS to accomplish a goal
- how to effectively apply SAS to your community or job
- how users like you implemented SAS to solve their problems

You, too, will be able to harness The Power of SAS!

This book is broken down into easy-to-read chapters that introduce you to SAS vocabulary and structure, show you how to plan and execute a successful project, introduce you to simple statistics, and walk you through a few case studies to help inspire and motivate you to use analytics for a real project. At the end of each chapter, there are five quick tips to take away from the content learned. After reading this book, you will be ready to plan, create, and execute a project using SAS.

Assembling Casseroles and Analyzing Data

"More than just something to eat, more than just a dish on the table, casseroles serve a purpose; they elicit a response. They are not an ingredient; casseroles are a genre." Vivian Howard, A Chef's Life

I am a North Carolina girl, born and raised in Charlotte. I had a wonderful childhood split between "city" life and a more rural atmosphere, and we loved to eat! Food is a theme that runs through this book, and I often use it as a metaphor in the classroom. My grandparents and parents came from a generation that ate three meals a day at the kitchen table, which, in turn, is much how I grew up. My Papa always said that food brings fellowship, and he is correct. Food is something that we can all relate to and it gives us common ground!

Figure 1.1 *Green Bean Casserole*

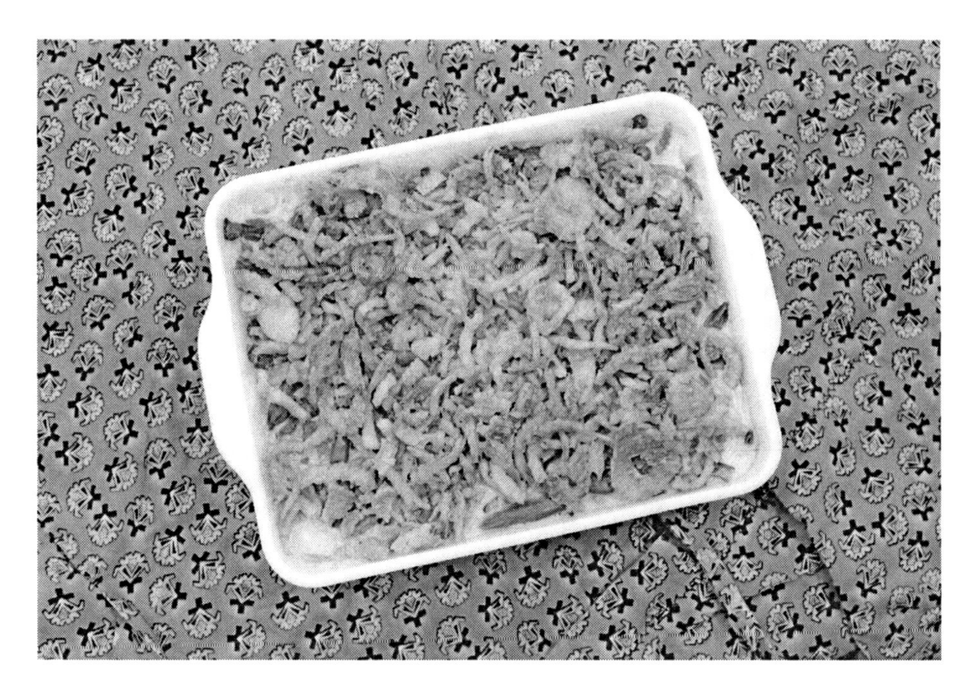

"The goal of a casserole is to feed the hungry and the heartbroken; something warm and predicable hits the spot." Sheri Castle, Southern writer

Why are casseroles a genre? Put very simply, they are made to feed the hungry, and they include simple yet hearty ingredients. Casseroles take different elements to create one entity.

So what does this have to do with data? I think about data like I think about food. If you watch a cooking show, the host always follows a plan to create a dish. There is a beginning, middle, and end to the recipe process. Working with data has the same concept. When you prepare a casserole, you put all of the ingredients together in a pan and then you put it in the oven. When you work with data, your "ingredients" are variables and observations. When you create a data analytical program, you put all of the variables and observations together in the program and then you run it.

You probably have a few questions. What is data analytics? Why is data analytics important? These are a few of the questions that have probably driven you to explore SAS. The terms "data" and "analytics" are more common parts of the social vernacular, and you have likely heard the term "data-driven decisions" in different contexts. We are producing more data than ever before with our plugged-in lifestyles. As a result, questions arise about data and how it can be used. Understanding what to do with data is only half the problem, and even just starting with basic analytical practices can be a pivotal change.

We can relate data analytics to a casserole because we understand the simplicity and the depth of a good casserole. Think about your favorite casserole. What image comes to mind? My mind immediately sees a green bean casserole. What are the key ingredients of your favorite casserole dish? Sheri Castle says that, *"If you make that casserole too "dolled" up, then you have missed the target. You may have hit another target, but your main target has been missed."* When baking a casserole or analyzing data, the main target is to create a simple but pragmatic entity from key ingredients.

For good data analytics, we want many people to be able to use and comprehend the information that we compile. Like a casserole, we want simple ingredients to feed many people.

Figure 1.2 *Ingredients and Data*

Figure 1.3 *Data to Analysis*

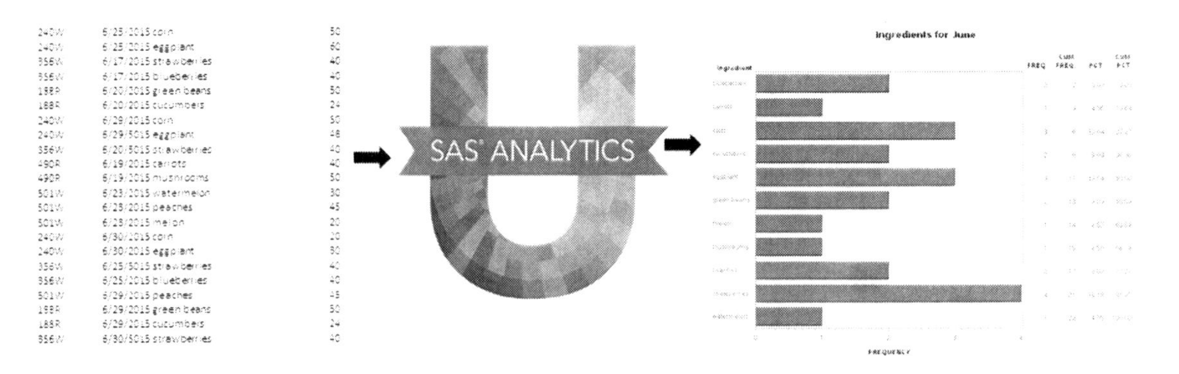

Keep thinking about the concept of a casserole and let's look at data. Data is everywhere, shedding light on all aspects of life. Retailers know what's selling and who's buying. Polls test opinions on everything from political candidates to consumer goods. Doctors monitor their patients' vital signs and progress over time. Social networks register the interactions of millions. At sporting events, fans and coaches examine collected comprehensive statistics on their teams' performance. If something can be measured, then in all likelihood, a vast archive of data is already being compiled. Often data is unprocessed, waiting for someone to analyze it and discover new and valuable knowledge about the world.

So, exactly what is data analytics? Analytics can be defined in many different aspects. In short, analytics create insight. Simply put, someone wants an answer, so a question is asked. To properly phrase the question and find the answer, we look at what we know, how we know it, and what more could we know about it? With SAS, you have a tool that can refine those questions and return easy-to-understand results, and those results create insight and uncover knowledge.

Where do you get data? To start, you can do something as simple as writing down what you eat for lunch everyday and each item's calories. Or, you might be interested in the top television shows, music, or movies. If so, you can find this data on the Nielson site, download the Nielson ratings data, and upload the data to SAS. If you want to learn something about your community, you can work with your local government to research local event data or see how a non-profit delegates funding. Maybe you want to know how many Girl Scout Cookies your troop sold over the past two years, or how much popcorn the Boy Scouts sold. Or, maybe you just want to analyze your spending habits to help yourself start saving money. It is all about the data!

SAS Software

To start analyzing data and getting results, download SAS University Edition. SAS University Edition is free SAS software that can be used by anyone who wants to learn SAS! With this software, you can explore and solve important and stimulating problems. Gaining new perspective on the power of data and the insight provided through reliable analysis can open your eyes to many new possibilities and solutions to various problems. SAS University Edition provides the opportunity for you to expand your knowledge of statistics and quantitative methods by offering faster and easier access to the most up-to-date statistical methods. Writing and submitting code in SAS University Edition is easy. It has a powerful graphical user interface where there is a multitude of opportunities to run analyses.

The user interface for SAS University Edition, also known as SAS Studio, is a browser interface. It can be accessed from common Internet browsers including Safari, Firefox, Internet Explorer, and Chrome. The SAS University Edition program is robust and offers analytic capabilities from basic to advanced. Below is an overview of the key features of SAS University Edition. There are five significant features that help a user learn and use analytics with SAS. These include:

1. **Base SAS:** Make programming fast, easy, and graphical with the Base SAS programming language, ODS graphics, and reporting procedures.

2. **SAS/STAT:** Choose from a variety of statistical methods and techniques in the Tasks and Snippets sections.

3. **SAS/IML:** Use this matrix programming language for more specialized analyses and data exploration.

4. **SAS Studio:** Reduce programming time with autocomplete for hundreds of SAS statements and procedures and build in syntax.

5. **SAS/ACCESS:** Connect with data, no matter where it resides.

As you begin to work in SAS University Edition, you will find that you can use all or just a few of these features to help you find the answers that you need. In this book, we focus on understanding Base SAS and using Base SAS in SAS Studio.

Starting SAS University Edition

Now that you know why you should use SAS to learn analytics, let's get started with the software! The direct link to download SAS University Edition is:

http://www.sas.com/universityedition

For more information about how to get your software up and running and how to access resources like FAQs and video tutorials, visit the Getting Started page on the SAS support site at:

http://support.sas.com/software/products/university-edition

Windows and Commands in SAS University Edition

Using the Windows in SAS University Edition

Once you open the software, you'll see that SAS University Edition has two basic windows. The Navigation pane houses the files, folders, tasks, snippets, libraries, and file shortcuts. The Work area contains four tabs: Program, CODE, LOG, and RESULTS.

Figure 1.4 *SAS University Edition Snapshot*

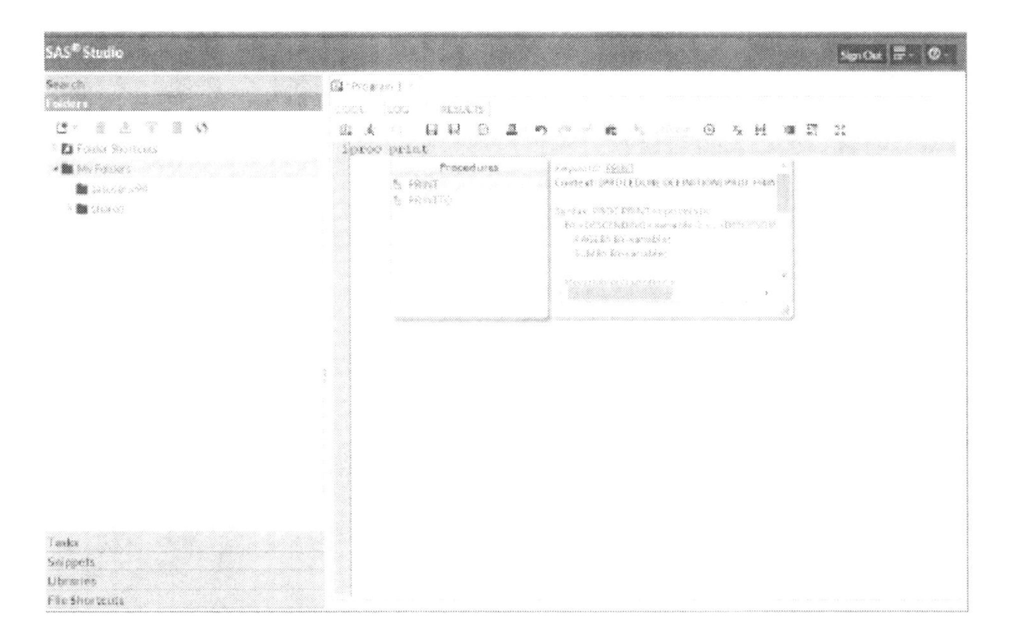

Navigation Pane

The Navigation pane gives you access to files, folders, tasks, snippets, libraries, and file shortcuts. Expand each item to show its contents.

- Server Files and Folders: Displays the content that is visible to SAS University Edition.
- Tasks: Contains menus for data manipulation, graphics, and statistical analysis.
- Snippets: Contains saved segments of SAS syntax or code that can be edited. You can also add your own snippets.
- Libraries: Contains a collection of SAS files. (SAS files are stored in libraries.)
- File Shortcuts: Provides shortcuts similar to what's in the Microsoft environment.

Figure 1.5 Navigation Pane

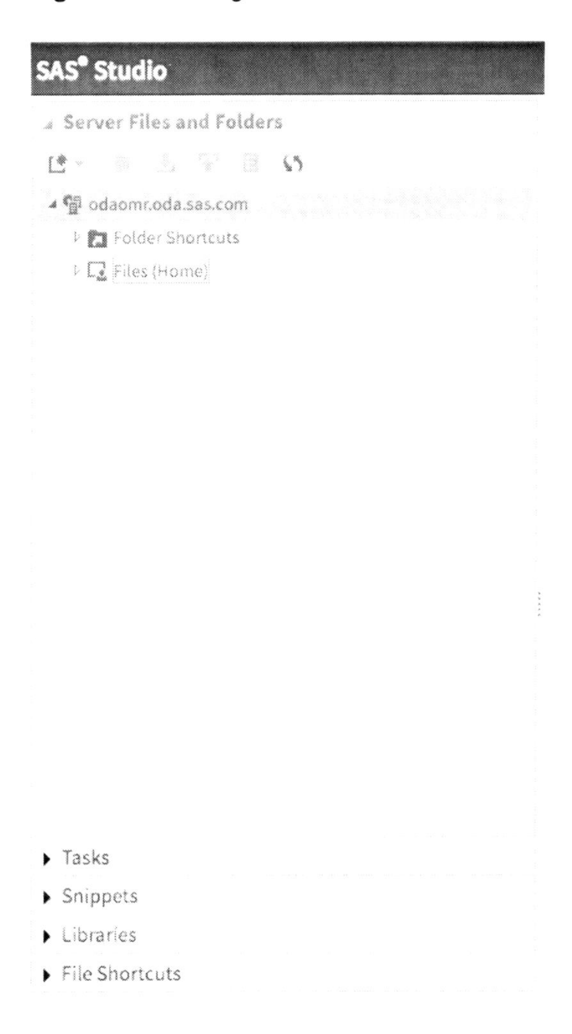

Work Area

The largest area of the screen is the Work area, where you draft a program or implement a task. The Work area has tabs that display different elements of your program or task.

Program: This tab displays the active program.

CODE: This tab is where you type, edit, and submit your program.

LOG: This tab contains notes about your SAS session. After you submit your program, it shows any notes, errors, or warnings. These items are color-coded.

RESULTS: This tab shows your output and generates any printable results.

Figure 1.6 *Work Area*

Drop-down menus and icons on the toolbar enable you to perform a variety of tasks, including submitting your program; opening a program; saving files; printing, cutting, and pasting text; and undoing or redoing an action.

Figure 1.7 *SAS Studio Toolbar*

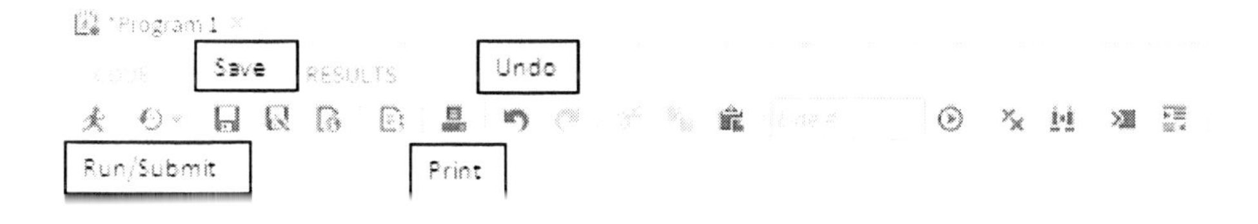

Now that you have an idea of the free software that you can download and use to do analytics, let's dig a little deeper into the SAS language and environment. The next few sections introduce you to the SAS language and its associated syntax.

You can find more resources about SAS University Edition through SAS free e-Learning at http://support.sas.com/edu/elearning. You can get started with SAS tutorials at http://support.sas.com/training/tutorial.

SAS Language

The SAS language is a computer programming language used for statistical analysis. SAS software can read data from common spreadsheets or databases. Or, you can create your own data in SAS and output it to tables and graphs. Most software applications are menu-driven or command-driven. This means that either a menu drives your coding or commands that you enter get results. SAS is actually neither. SAS uses statements to write a series of instructions called a program. The program communicates what you want to do and is written using SAS language syntax. If you want, you can do menu-driven coding in SAS University Edition. However, you have more flexibility and get more control over your code if you learn how to write your own program. Understanding how syntax is written and executed helps you better manipulate your data.

SAS Programs

A SAS program is a sequence of statements that execute one step at a time, statement by statement, independently of one another. Whereas other programming languages compile an entire program and then completely execute it, SAS reads each line of information separately. To better imagine how a SAS program works, think about your favorite restaurant. You enter the restaurant, take a seat, and then when the waiter or waitress arrives, you say what you would like to eat. When you state your order, think of the statements like a program:

```
I would like to place my order for my meal. I would like a glass of
water. I would like to order the chicken sandwich and chips. I would
also like a napkin.
```

Note that you first say what you want to do. You add subsequent statements that further detail your request. For example, in a program, you might want to know how much money was made in a quarter, or what was the percentage of individuals participating in a sport? Within a SAS program, SAS statements are what make the program run.

SAS Statements

SAS programs are made up of steps, and steps are made up of statements. Remember, you are learning a new language, and when you create a SAS program, as with any programming language, there are rules to follow. Your SAS program is constructed of DATA and PROC steps. *DATA steps* are typically used to create SAS data sets. *PROC steps* (PROCedures) are used to process SAS data sets (or, in other words, to generate reports and sort data). The most important rule is:

Every SAS statement ends with a semicolon.

Although this might seem simple, even the most experienced programmer will forget a semicolon! In the following programming example, PROC PRINT is used to display the data in a table format on the RESULTS tab. The data equals (data =) option references the data set that you would like to see.

```
PROC PRINT data = myfolder.inventory;
Run;
```

Figure 1.8 on page 11 shows a program in SAS Studio. Notice that each statement ends with a semicolon.

Figure 1.8 *SAS Coding Environment*

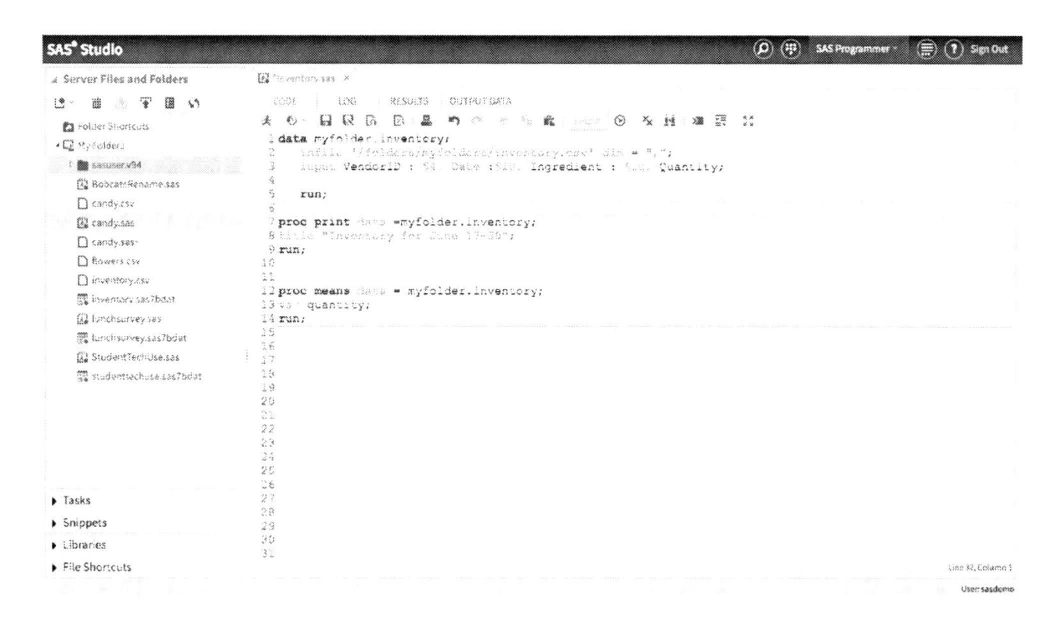

The DATA step portion creates a permanent or temporary SAS data set from the raw data or from another SAS data set. SAS procedures can be run only on SAS data sets. A temporary SAS data set is a data set that is created in the program and deleted when the SAS Studio session ends. A permanent SAS data set is useful once you have gotten your data in the format that you need and you expect to be working with it repeatedly.

As you are programming, you will be working on three main tabs of the graphical user interface:

- CODE tab – where you write your code
- LOG tab – where a transcript of your code is shown

- RESULTS tab – where code results are shown

SAS has color-coding to help you spot errors as you are programming. There is a predominant blue color scheme. Dark blue is used for DATA and PROC steps. Light blue is used for statements within your steps. And, purple is used for quotations.

When you run your SAS program, the log shows the processing that occurred and also uses colors. The log uses three main colors: red, blue, and green. Errors are in red, warnings are in green, and notes are in blue. The color-coding in SAS helps you see whether there was a problem, where the problem occurred, and how to fix the problem.

Figure 1.9 *SAS Log Color-Coding*

*Inventory.sas ×

CODE LOG RESULTS

Errors, Warnings, Notes

Errors (1)

ERROR: Unrecognized form of the RUN statement. Use either RUN; or RUN CANCEL;.

Warnings (1)

WARNING 1-322: Assuming the symbol DATA was misspelled as dat.

Notes (7)

```
      real time           0.01 seconds
      cpu time            0.01 seconds

67         !                            QUIT;
68           QUIT;RUN;
69           ODS HTML5 (ID=WEB) CLOSE;
70
71           ODS RTF (ID=WEB) CLOSE;
72           ODS PDF (ID=WEB) CLOSE;
NOTE: ODS PDF(WEB) printed 1 page to /tmp/SAS_workD30E00000A0E_loca
73           FILENAME _GSFNAME;
NOTE: Fileref _GSFNAME has been deassigned.


74           DATA _NULL_;
75           RUN;

NOTE: DATA statement used (Total process time):
      real time           0.00 seconds
      cpu time            0.00 seconds
```

Layout of a SAS Program

When creating a SAS program, the layout is flexible. It is good programming practice, however, to format your program in a neat-looking fashion so that it is easy to read and execute. Remember that a SAS program is made up of SAS statements and SAS statements analyze data. The SAS program is the shell that holds the statements. You should avoid writing code that is difficult to read. Here is an example of good layout:

```
Data myfolder.inventory;
    infile "/folders/myfolders/garden.dat";
    input vegetables $ fruit $;
Run;

PROC PRINT data=myfolder.inventory;
Run;
```

In the code, there are several SAS statements, each ending with a semicolon. In addition to the code, you can add comments. Comments make your program more understandable, and you can insert comments anywhere in the program that you like. SAS does not read comments as code. Comments are usually used to annotate your program, making it easier for someone to read and understand what you have done and why you have done it. The syntax for a comment is /* to initialize and */ to close. When you are programming in SAS, your comment code is color-coded green.

Here is an example of how to comment your code:

```
/*Creating a data set from the raw data file garden.dat*/

Data myfolder.inventory;
    infile "/folders/myfolders/garden.dat";
    input vegetables $ fruit $;
Run;

PROC PRINT data=myfolder.inventory; /*Printing the data in the created
    data set myfolder.inventory*/
Run;
```

TIP When you are first learning a programming language, it can be frustrating. It is much like learning a foreign language with all the rules and exceptions. The first time that you code a program, it will probably not run. That is okay! We all learn from our mistakes. Sometimes we can get caught up in the idea that it has to run perfectly the first time we write the code, but that is just not realistic. Many times it is because you forgot a semicolon. (Tradeoff: You get to join the Semicolon Club! I have been a member for many years!) When you are first learning the programming language, write simple programs. Do not try to tackle complicated multi-step programs until you are comfortable. Always check your log after your program runs to confirm the results. As you get into that habit, you will increase your efficiency and begin to have more confidence in your coding. Remember, build a program piece by piece. This allows for easy-to-maintain code that is manageable should a problem arise.

SAS Data Sets

Useful Vocabulary

To run analysis on data that you have collected, the data must be in a format that SAS can read. Therefore, your data must be in the form of a SAS data set. A SAS data set is a specially structured file that contains data values. SAS data sets are processed by SAS procedures. SAS has a robust ability to work with and read almost any data. Once the observations or files are in SAS, the data is stored and can be opened whenever needed to process analyses and reports. A few vocabulary words that are associated with a data set include:

- variables
- observations
- descriptor portion
- data portion
- data types
  - numeric variables
  - character variables
- missing data

Variables and Observations

To better understand a SAS data set, you must understand SAS terminology. Data is the primary component of the data set, and every language has terminology that is associated with that data. In SAS, the terminology is:

- SAS data sets are also called tables.
- Observations are also called rows.
- Variables are also called columns.

Figure 1.10 *SAS Data Set Terminology*

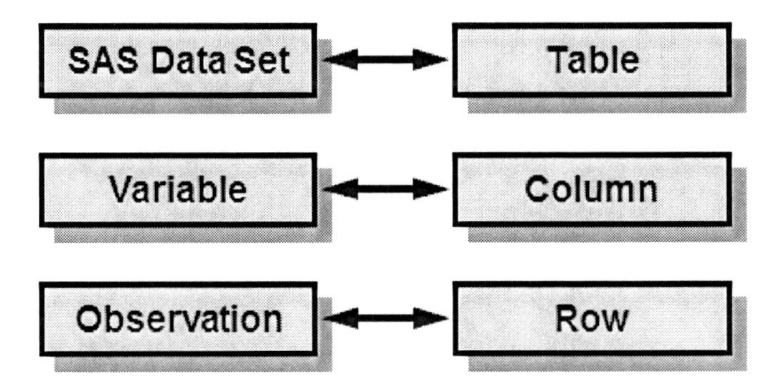

In Table 1.1 on page 15, you see how variables are columns and observations are rows. The abbreviation "Obs" stands for observations.

Table 1.1 *Student Data Set, Variables (Columns) and Observations (Rows)*

Obs	Name	Age	Gender	Grade
1	Lauren	17	F	Senior
2	Jonathan	16	M	Junior
3	Joshua	17	M	Senior
4	Rachel	17	F	Senior
5	Jacob	16	M	Junior

Figure 1.11 *Output in SAS*

Student Information

Obs	name	age	gender	grade
1	Lauren	17	F	Senior
2	Jonathan	16	M	Junior
3	Joshua	17	M	Senior
4	Rachel	17	F	Senior
5	Jacob	16	M	Junior

To further explore SAS terminology, let's look at the descriptor portion and the data portion of a SAS data set.

Browsing the Descriptor Portion

The descriptor portion of a SAS data set simply *describes* the data. It contains general information about the SAS data set such as the data set name and number of observations. This portion of the data set includes variable attributes, such as variable name, type (character or numeric), length, format, and label. You can see the descriptor portion of the data set using the CONTENTS procedure (PROC CONTENTS).

```
PROC CONTENTS data = myfolder.inventory;
Run;
```

Figure 1.12 *PROC CONTENTS*

The CONTENTS Procedure

Data Set Name	WORK.INVENTORY	Observations	22
Member Type	DATA	Variables	4
Engine	V9	Indexes	0
Created	11/02/2015 12:51:58	Observation Length	48
Last Modified	11/02/2015 12:51:58	Deleted Observations	0
Protection		Compressed	NO
Data Set Type		Sorted	NO
Label			
Data Representation	SOLARIS_X86_64, LINUX_X86_64, ALPHA_TRU64, LINUX_IA64		
Encoding	utf-8 Unicode (UTF-8)		

Engine/Host Dependent Information	
Data Set Page Size	131072
Number of Data Set Pages	1
First Data Page	1
Max Obs per Page	2722
Obs in First Data Page	22
Number of Data Set Repairs	0
Filename	/saswork/SAS_work27450000F02F_odaws03-prod-us/SAS_work14060000F02F_odaws03-prod-us/inventory.sas7bdat
Release Created	9.0401M3
Host Created	Linux
Inode Number	1046593
Access Permission	rw-rw-r--
Owner Name	sharon.jones0
File Size	256KB
File Size (bytes)	262144

Alphabetic List of Variables and Attributes			
#	Variable	Type	Len
2	Date	Char	10
3	Ingredient	Char	20
4	Quantity	Num	8
1	VendorID	Char	4

The descriptor portion of your data set is extremely valuable as you become familiar with the data. It displays the variable attributes that help you better understand your data and how you would like to proceed with analyses.

Understanding the Data Portion

Once your data has become a SAS data set, you can see the listing of the observations in the Output Data window.

Figure 1.13 *Seeing Data in the Output Tab*

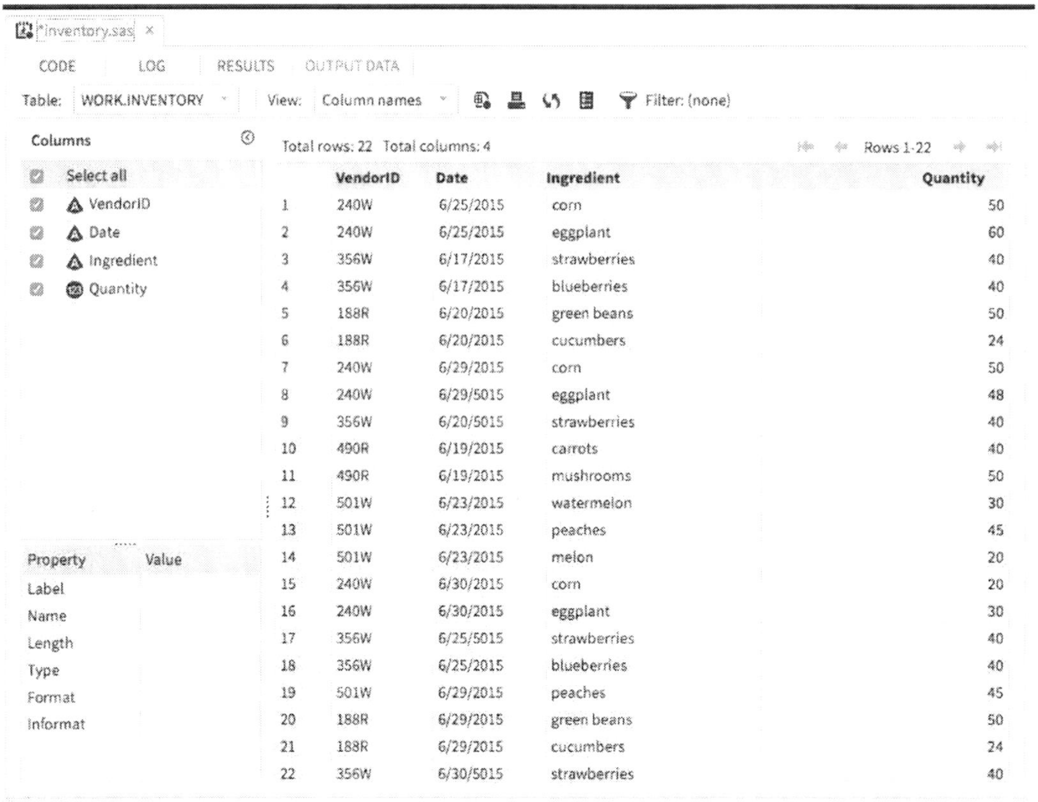

*inventory.sas ×

CODE | LOG | RESULTS | OUTPUT DATA

Table: WORK.INVENTORY View: Column names Filter: (none)

Columns

- Select all
- VendorID
- Date
- Ingredient
- Quantity

Property	Value
Label	
Name	
Length	
Type	
Format	
Informat	

Total rows: 22 Total columns: 4 Rows 1-22

	VendorID	Date	Ingredient	Quantity
1	240W	6/25/2015	corn	50
2	240W	6/25/2015	eggplant	60
3	356W	6/17/2015	strawberries	40
4	356W	6/17/2015	blueberries	40
5	188R	6/20/2015	green beans	50
6	188R	6/20/2015	cucumbers	24
7	240W	6/29/2015	corn	50
8	240W	6/29/5015	eggplant	48
9	356W	6/20/5015	strawberries	40
10	490R	6/19/2015	carrots	40
11	490R	6/19/2015	mushrooms	50
12	501W	6/23/2015	watermelon	30
13	501W	6/23/2015	peaches	45
14	501W	6/23/2015	melon	20
15	240W	6/30/2015	corn	20
16	240W	6/30/2015	eggplant	30
17	356W	6/25/5015	strawberries	40
18	356W	6/25/2015	blueberries	40
19	501W	6/29/2015	peaches	45
20	188R	6/29/2015	green beans	50
21	188R	6/29/2015	cucumbers	24
22	356W	6/30/5015	strawberries	40

The Output Data tab holds the data. You can open the table and view the data in SAS. To view the table, click **Libraries** in the Navigation pane, click a folder, and then click a data set. Figure 1.14 on page 19 shows these steps. The data portion is the actual list data; the descriptor portion describes the data.

Figure 1.14 Viewing the Data Table in SAS Studio

	VendorID	Date	Ingredient	Quantity
1	240W	6/25/2015	corn	50
2	240W	6/25/2015	eggplant	60
3	356W	6/17/2015	strawberries	40
4	356W	6/17/2015	blueberries	40
5	188R	6/20/2015	green beans	60
6	188R	6/20/2015	cucumbers	24
7	240W	6/29/2015	corn	50
8	240W	6/29/2015	eggplant	48
9	356W	6/20/2015	strawberries	40
10	490R	6/19/2015	carrots	40
11	490R	6/19/2015	mushrooms	50
12	501W	6/23/2015	watermelon	30
13	501W	6/23/2015	peaches	45
14	501W	6/23/2015	melon	20
15	240W	6/30/2015	corn	20
16	240W	6/30/2015	eggplant	30
17	356W	6/25/2015	strawberries	40
18	356W	6/25/2015	blueberries	40
19	501W	6/29/2015	peaches	45
20	188R	6/29/2015	green beans	50
21	188R	6/29/2015	cucumbers	24
22	356W	6/30/2015	strawberries	40

Data Types

Figure 1.15 Character and Numeric Variables

In SAS, there are two data types—numeric and character. Statisticians often refer to numeric variables as quantitative variables and character variables as qualitative variables. Numeric or quantitative variables are numbers that can be positive or negative and added or subtracted. In addition, they can be decimals. Character or qualitative variables contain any value including letters, numbers, special characters, and blanks.

A numeric variable can have a plus (+), minus (-), decimal (.), or E for scientific notation. These are all valid numeric options. The default storage or length of a numeric variable is 8 bytes, but it is not restricted to 8 digits. The value of a numeric variable is right-aligned when it is displayed in output.

Figure 1.16 *Right-Aligned Numeric or Quantitative Variables Output in SAS*

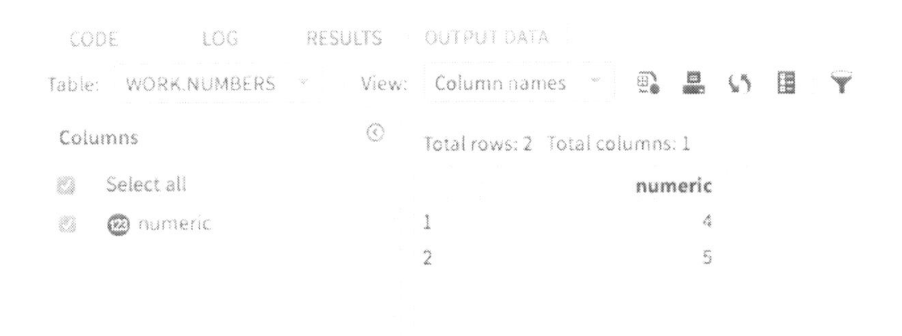

A character variable can have any value including letters, numbers, special characters, and blanks. If a variable contains letters or special characters, then it must be a character variable. If a variable contains only numbers, then it can be either numeric or character, depending on the use of the variable in the data set. Some numbers might make more sense as a character variable, like a phone number or employee ID, because you would not use mathematics on these variables. A character variable is left aligned.

Figure 1.17 *Left-Aligned Character or Qualitative Variables Output in SAS*

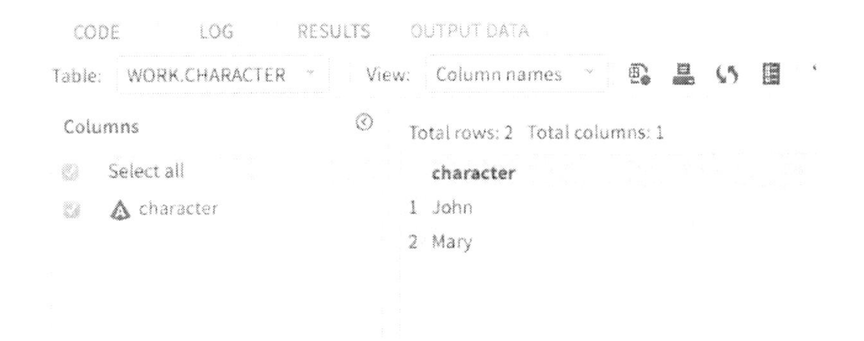

Variables are the columns of the data set and the pivotal part of the analysis. When naming a SAS variable, there are a few rules to keep in mind:

- A name must be 32 characters or fewer in length.
- A name must start with a letter or an underscore. It cannot begin with a numeral or contain special characters.

Names should be relevant and simple. When working with variables in your SAS data set, it is much easier to manipulate and analyze variables that have simple names. For example:

Favorite Color = Favcolor

Last Name = LastName

Simple relevant names make programming easier. Later on, you can label your variable to be more complex for a presentation.

Missing Data

When working with data, it is often common to have missing values. Missing data can still be read in SAS, though. A missing value is noted with a period (.) for numeric data and with a blank space for character data. In Table 1.2 on page 21, the input data did not include an age for John, and there was no value for Name in the second row. The period in the first observation represents a missing numeric value, and the blank space in the second row represents a missing character value.

Table 1.2 *Missing Data in Numeric and Character Variables*

Obs	Age	Name
1	.	John
2	8	

Figure 1.18 *Missing Data Output in SAS*

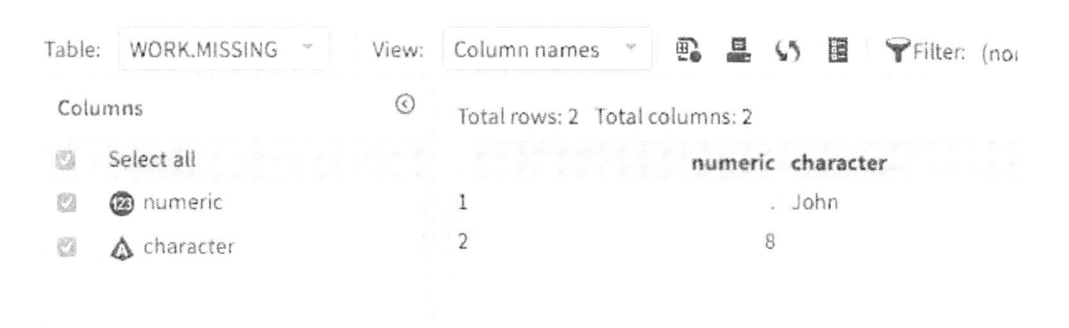

Don't worry, missing data is a normal part of any data project and it can be accounted for in the SAS system.

Additional Resources

As you begin working in SAS University Edition, you might have questions or thoughts to share. SAS has an online community that can help answer those questions and connect you with other programmers! You can visit the SAS Analytics U site at http://communities.sas.com/t5/SAS-Analytics-U/bd-p/sas_analytics_u and post questions, review other questions asked, or just chat with other users. It is a great resource and is always available!

If you prefer a more hands-on option or a book, Ron Cody's *An Introduction to SAS University Edition* is very helpful.

Quick Tips

1. A SAS program contains one or more steps, and each step is a sequence of statements.
2. Every SAS statement ends with a semicolon.
3. Your SAS program is constructed of DATA and PROC steps.
4. In SAS, there are two data types or variable categories—numeric (quantitative), or character (qualitative).
5. Missing data can be read in SAS. A missing value is noted with a period (.) for numeric data and with a blank space for character data.

2

Planning and Executing a Successful Programming Project

Successful Planning

A Recipe for Success

Just like following a recipe for a casserole, creating a plan for a programming project is non-negotiable. Starting with good ingredients or with a good plan creates an ease of execution because you know where to begin and end. Sheri Castle states, "If you don't put something good in the dish, then you are not going to get anything good out of the dish; you can't just put whatever in there because nothing magic is going to happen when you are not looking. When it is in the oven, it is not going to get better if it did not go in there that way!"

When you are thinking about how to begin work on a project, there are a few steps that can help guide your planning process. This chapter walks through these steps (or ingredients) to include in your project plan (or recipe).

1 **Develop and Brainstorm** - Is there a question that has been asked or is there something you would like to know more about?

2 **Set Objectives and Goals** - What do you want to solve in your project?

3 **Identify Deliverables** - Give yourself reasonable deliverables to obtain. What will you be able to show when your project is completed?

4 **Plan a Schedule** - When do you want the project to be completed?

5 **Make Supporting Plans** - Identify resources that you need.

6 **Determine Applicable Code** - What code will you need to complete your project?

7 **Have Good Coding Habits** - How will you implement efficient code?

8 **Communicate** - Communicate with others on the project and be able to present the project to persons outside of the organization.

Figure 2.1 *Project Planning Visual*

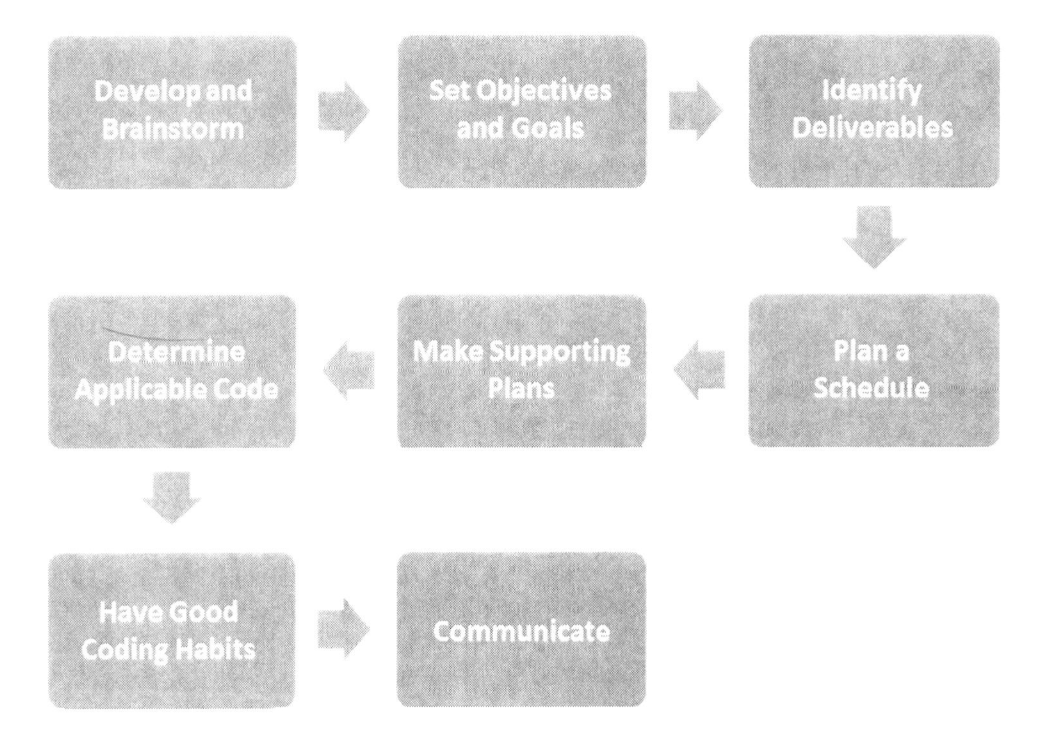

Eight Phases of Project Planning

Develop and Brainstorm

The development phase begins with a problem or question. In this phase, brainstorming is the catalyst for organization.

The following recipe is from my grandmother, Mama D., as I called her (short for Mama Dorothy). It is her "famous" strawberry shortcakes. I thought these were amazing as a kid, and they still are today!

Figure 2.2 *Strawberry Shortcake Recipe, Dorothy Moore*

Start brainstorming about what "recipe" you would like to make. What data do you have? What results are you looking for?

Some brainstorming questions to ask are:

- What is currently going on in our society?
- Is there a problem in your neighborhood, school, or city that needs a solution?
- Is there a pattern that you'd like to find?

Innovation has been thriving in terms of solving problems and answering questions for thousands of years. Think about how the light bulb or stove was created. Research shows that the wheel was the best invention of all time. The wheel is a simple concept, but it changed how humans were able to move. Now, that's some data!

This is the time to brainstorm. Think about what the project could include and how you might solve the problem or answer the question at hand.

- Will there be road bumps?
- Are there key stakeholders who need to be involved in the planning and execution of the project?

During this phase, there are no right or wrong thoughts. This is the time to explore ideas and find what answers the question and accomplishes the goal.

You need to think about where you will get your data.

- Will you be collecting the data yourself?
- Will you be receiving data from a third party?

Where you get your data plays a big part in how you execute the remaining steps of the planning process. Many times, collecting your own data can take a bit longer than receiving data from a third party, and neither way is better than the other. Collecting your own data can be rewarding because it gives you control over the content.

Set Objectives and Goals

When you know the question that you want to answer, how do you start your project? Think about how you start a recipe—you start with the ingredients. In the following photo, the ingredients have been collected and laid out before assembling the recipe. Similarly, you want to define a clear goal and plan for your project before you start "cooking."

Figure 2.3 *Ingredients for Strawberry Shortcake*

The recipe provides an outline of what goes in your dish. Each ingredient, just like each part of the project planning process, has a purpose and is used to move the process forward.

Your objectives are the specifics that your project should accomplish. These specifics are the measurable results that, when reached, mean that your goal has been achieved. For example, maybe you want to contact all the youth in your school or neighborhood to make them aware of crime problems and to get two-thirds of them to join a crime watch program. Finding the measurable goal is detailed in the *scope statement*. The *scope statement* is unarguably the most important part of the project plan. The scope statement is the common agreement among all stakeholders about the project's definition.

Identify Deliverables

A deliverable is a tangible and measurable result, outcome, or product that completes a project or part of a project. Typically, the stakeholders agree on deliverables to ensure that the project meets expectations. A deliverable can be a presentation, a report, a prototype, a visual graph, or any other tangible item that represents your work. Remember, the reason that the project was created was because there was a question to answer.

Plan a Schedule

Once you have an idea about your project, its objectives, your goals, and how you are going to deliver the results, you need a schedule. Just like making strawberry shortcake, a project takes time. The shortcakes cook in the oven for 12 to 15 minutes. Your project is on a schedule as well. Beware, as you work through the project and it begins to take shape, it might make some unintended turns.

Is this a project that you can complete in a day or in a week? Is it a multi-week project? Once you know how long the project will take, a schedule will help keep the project on track. One way to stay on track is to create markers on your calendar that remind you to schedule weekly goals and communicate updates. Establishing a beginning, middle, and end creates a structure and provides an outline that will help see your project through to the end. In addition, having checkpoints holds you and others working on the project accountable for your deadlines.

Make Supporting Plans

In addition to the analytical software that uses the data, is there other software required to execute your project? Are there academic resources that are needed? How about cloud resources? Or, graphic design capabilities? When planning your project, you want to make sure that you have all the necessary resources to execute your project with minimal interruption.

Determine Applicable Code

Now it's time to put it all together. In Figure 2.4 on page 28, the ingredients have been combined and they are ready to go in the oven. Regarding your code, you have your variables defined and the questions to answer laid out. Your project is ready to move into "applicable code mode."

Figure 2.4 *Putting Ingredients Together*

As you use SAS University Edition and become comfortable with the SAS language format and syntax, you will be able to make decisions about the applicable code for your project. There are several PROC statements that are staples for understanding data:

- PROC CONTENTS explains the contents of the data set.
- PROC PRINT provides a printable view of the data.
- PROC SORT sorts data based on a variable either in ascending or descending order.
- PROC MEANS provides summary data on numeric variables.
- PROC FREQ produces frequency counts and crosstabulation tables and shows relationships among variables.
- PROC CORR displays correlations between variables.

Other basic analytical procedures can be used to understand data, and they are discussed in Chapter 3. Here is an example of a PROC step. PROC MEANS shows the average quantity of ingredients in pounds bought by a chef for her restaurant, the standard deviation, and the minimum and maximum.

```
PROC MEANS data=myfolder.inventory;
    var quantity;
```

```
run;
```

***Figure 2.5** PROC MEANS Code in SAS*

*Inventory.sas ×

CODE LOG RESULTS

```
1 proc means data=myfolder.inventory;
2     var quantity;
3 run;
4
5
6
7
8
9
10
```

In Figure 2.5 on page 29, the data set inventory is in the myfolder library. The program ran PROC MEANS on the variable QUANTITY.

***Figure 2.6** PROC MEANS Results in SAS*

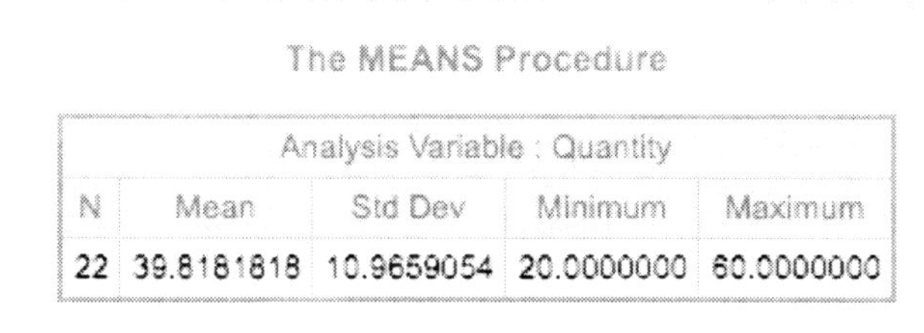

The MEANS Procedure

Analysis Variable : Quantity				
N	Mean	Std Dev	Minimum	Maximum
22	39.8181818	10.9659054	20.0000000	60.0000000

In Figure 2.6 on page 29, the variable QUANTITY, in pounds of ingredients, has an average, minimum, and maximum of 39.82, 20, and 60, respectively. This is a snapshot of a particular time frame for the restaurant.

Using the same data set, accessible in the Example Code and Data on the SAS Press author page, let's look at a few frequencies. Use the inventory data set again, and look at the frequency of the type of ingredient bought and the frequency at which an ingredient was purchased from a vendor. The same setup that we used with PROC MEANS applies here. However, notice we use the keyword TABLES instead of the variable QUANTITY because PROC FREQ runs as a table for each variable designated.

```
PROC FREQ data=myfolder.inventory;
    tables ingredient vendorid;
run;
```

Figure 2.7 *PROC FREQ Code in SAS*

*Inventory.sas ×

CODE LOG RESULTS

```
1 proc freq data=myfolder.inventory;
2     tables ingredient vendorid;
3 run;
4
5
6
7
8
9
10
```

In Figure 2.7 on page 30, the data set inventory is in the myfolder library. The program ran PROC FREQ using the keyword TABLES and the variables INGREDIENT and VENDORID.

Figure 2.8 *PROC FREQ Results in SAS*

The FREQ Procedure

Ingredient	Frequency	Percent	Cumulative Frequency	Cumulative Percent
blueberries	2	9.09	2	9.09
carrots	1	4.55	3	13.64
corn	3	13.64	6	27.27
cucumbers	2	9.09	8	36.36
eggplant	3	13.64	11	50.00
green beans	2	9.09	13	59.09
melon	1	4.55	14	63.64
mushrooms	1	4.55	15	68.18
peaches	2	9.09	17	77.27
strawberries	4	18.18	21	95.45
watermelon	1	4.55	22	100.00

VendorID	Frequency	Percent	Cumulative Frequency	Cumulative Percent
188R	4	18.18	4	18.18
240W	6	27.27	10	45.45
356W	6	27.27	16	72.73
490R	2	9.09	18	81.82
501W	4	18.18	22	100.00

In Figure 2.8 on page 31, strawberries were purchased the most. Vendors 188R and 501W had products purchased from them more than others during the time frame.

Choosing the applicable code for your data set depends on your questions and answers. Are you looking for basic summary statistics or a more in-depth analysis?

Have Good Coding Habits

When working with code, having good coding habits makes the coding process easy to follow. Good habits include following the syntax rules of the language and inserting comments in the code. In SAS, using comments help guide your project. When writing comments, begin with /* and end with */. You can see an example of comment code in Chapter 1.

Comments are extremely helpful when you are working with a team or when you need to share results. When writing a comment, you should elaborate and include important details so that it can help you fix problems in the future. In addition, save your project frequently to prevent loss of code. Format your code so that it is visually easy to read and understand. In this way, problems are easier to detect and successful code is easier to replicate. If there is code that is not needed, remove it.

When you are creating your program, consistency is often neglected. Being consistent when formatting and naming gives the program a good flow and fluidity. If a user can clearly read and understand your code and logic, then the project has a life of its own and can change as needed.

Here is a quick breakdown of good coding habits:

- Use comments in your code.
- Save frequently.
- Format code so that it is visually easy to read.
- Be consistent.

Striving to maintain good coding habits makes planning, implementing, and communicating about your project a positive and rewarding experience.

Communicate

As you move through your project, it is imperative to maintain communication among the stakeholders. As the project evolves, everyone involved should be aware of the process. Creating a weekly email, texting, and posting to a discussion board are good communication techniques. In addition, using cloud technology for collaboration documents is a great way to have real-time collaboration if everyone is not physically in the same location.

Working through the planning phases enables you to have tangible touchpoints and a process to follow. As you follow a recipe, you work your way through the process of creating a complete dish. The same is true for a project. A successful project has a path to follow, a relevant purpose, and well-executed results.

Figure 2.9 *The Final Product*

Implementing a Project

Introduction

The project plan defines your project and provides the information that helps the audience fully understand the scope of the project. In the project plan, the problem or question is addressed and how you intend to complete the task is explained.

Get Data

So, where do you get your data? You can collect your own data, find data on the Internet, receive data from a third party, or have data already in a database. If you are collecting your own data, you can use a variety of methods including Excel, a survey tool, a Google form, or an online form.

In Table 2.1 on page 34, a simple Excel spreadsheet is used for data entry. Notice that the variable YEARSTEACHING is one word. SAS naming conventions recommend that a variable be one word. It must begin with an underscore or letter and it cannot contain blank spaces. Later on, you can use PROC PRINT to separate the words.

Table 2.1 *Excel Data Entry*

Name	Subject Taught	County	YearsTeaching

You can collect data using an online survey. There are a variety of survey tools on the market. Take some time to research and find the tool that you like.

Figure 2.10 *Survey Tool*

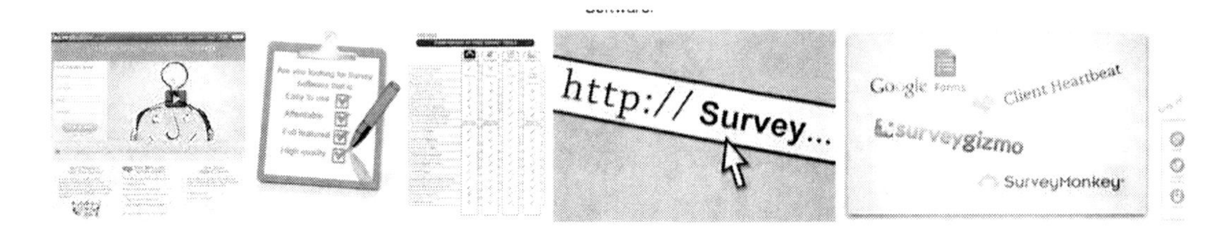

The following figure is an example of a Google form. The form can be set up through your Google account. It enables you to create questions with multiple choices, text, paragraph text, or a scale. Then, you present the live form as a link and your survey participants click on the link and answer questions.

Figure 2.11 *Google Form*

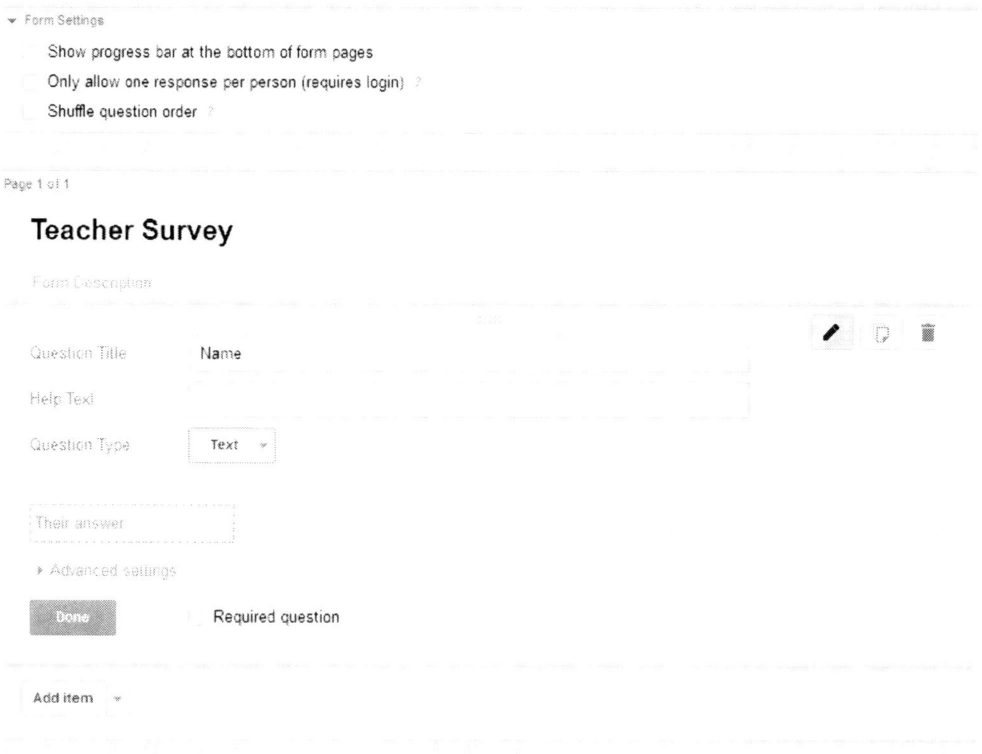

Form Settings
Show progress bar at the bottom of form pages
Only allow one response per person (requires login)
Shuffle question order

Page 1 of 1

Teacher Survey

Form Description

Question Title Name
Help Text
Question Type Text

Their answer

Advanced settings

Done Required question

Add item

Remember SAS naming conventions as you prepare your data for upload. SAS variables must begin with a letter or an underscore and they cannot contain blank spaces.

Accessing Data

Introduction

Once you have collected your data, you can begin the process of analysis. That is when you can use SAS! There are a few ways to import your data, but before you import it, you need to set up a library.

Common Methods

Data can come in many different forms and reside in various places. Wherever it is, it can be imported into SAS. As you collect or receive data for your project, be assured that you can access the data in SAS University Edition. The two most common methods for importing data into SAS are:

- Creating SAS data sets from raw data files.
- Converting another software's data into SAS data sets (using PROC IMPORT).

Figure 2.12 *Accessing Data Sources*

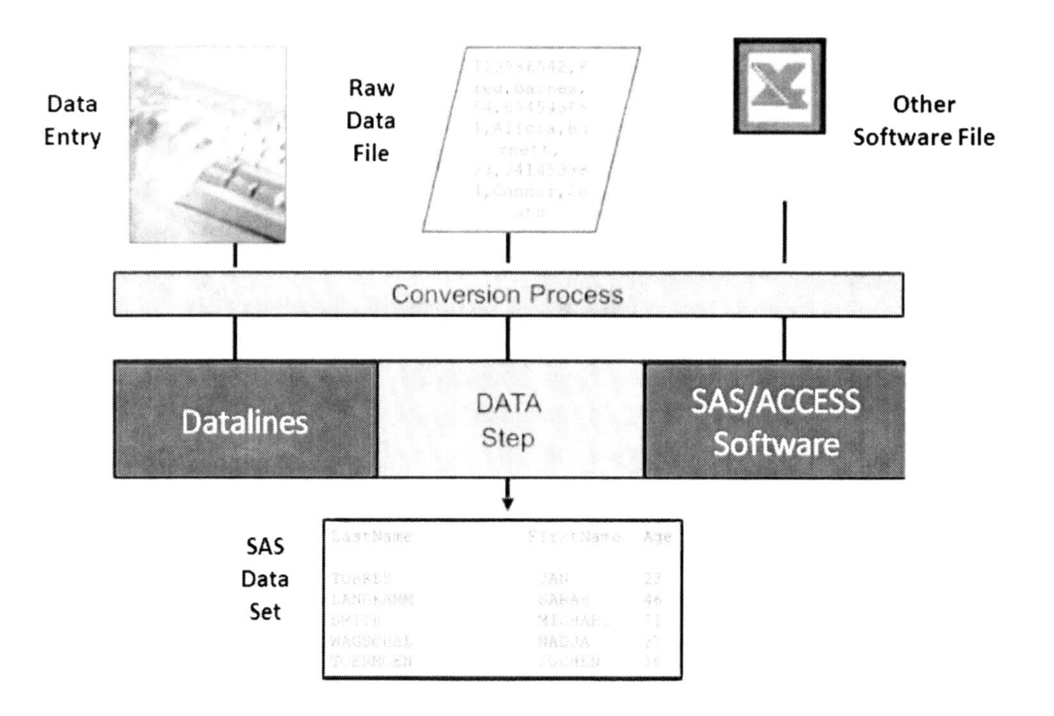

Before you begin accessing data, let's review the library setup. It is important that you upload your data to the library of pertinence in your project.

SAS Libraries

SAS references both the location and name of the data set. Typing the full path every time you use a data set is inefficient and time-consuming. SAS enables you to give a nickname to a fully qualified pathname. You can have a nickname for every different folder that contains SAS data.

SAS libraries are like a filing cabinet. All SAS files are stored in drawers, and each drawer is a different library. Libraries are simply directories that contain SAS files. More specifically, a library is a collection of one or more SAS files that are recognized by SAS and are referenced and stored in one location. You can have many different directories or folders that contain SAS files.

When you create a library, you assign it a libref (library reference). The libref is what you use to identify the library. You can create a permanent library or use the SAS temporary library, Work. The permanent library is present each time you launch a SAS session. A temporary library and its files are deleted when you end your SAS session. The temporary Work library can be used to store files while you are working in SAS. Figure 2.13 on page 37 has an example of the SAS Work library.

Figure 2.13 *SAS Work Library*

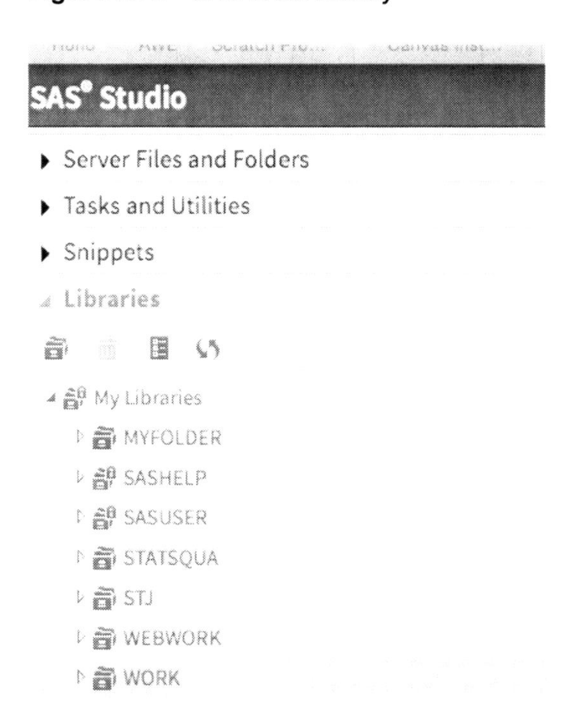

If you want to keep the library from one session to the next, you should create a permanent library. There are a few ways to do this in SAS Studio. You can create your own folder to reference as a library, or you can use the default shared folder to create a permanent library. The examples in this book use the default shared My Folders folder to store SAS data sets and libraries.

In SAS Studio, the default shared folder is **Server Files and Folders > My Folders**. In your operating environment, this shared folder is called Myfolders. If you have any existing SAS programs or data sets that you want to use in SAS University Edition, you can copy them to the Myfolders directory on your local computer. Whether the data is a SAS data set or a raw data set, it can be stored in Myfolders. You can then create a library in this folder in SAS.

In Figure 2.14 on page 38, the data that is in the **myfolders** folder on the desktop is the same data that is in SAS Studio. This enables you to always have access to your data in both.

Figure 2.14 *myfolders Content*

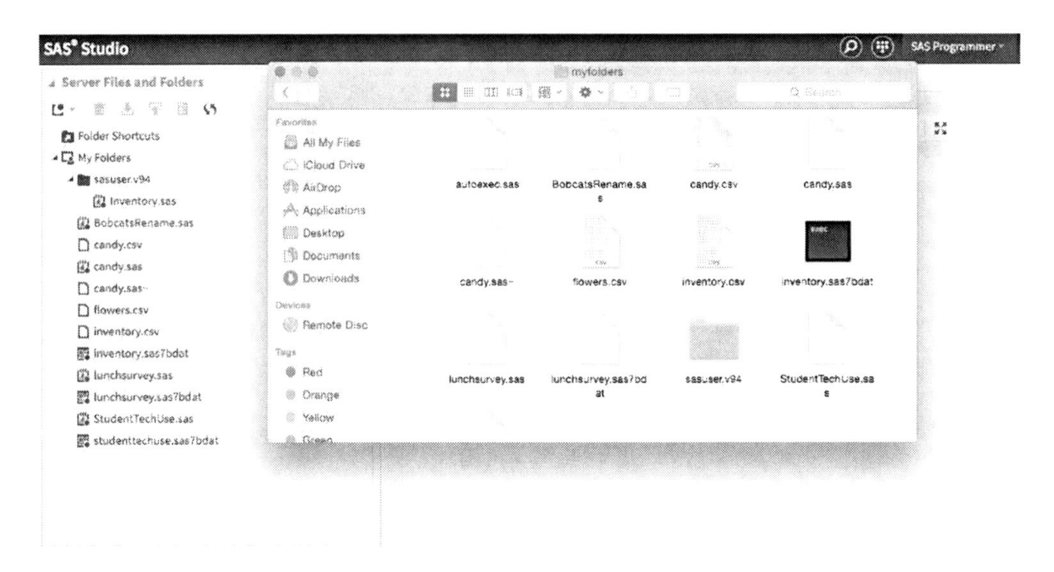

To create a permanent library from the shared folder:

- Under Server Files and Folders, select My Folders, and right-click.
- Select Create > Library.

Figure 2.15 *Creating a Library from a Shared Folder*

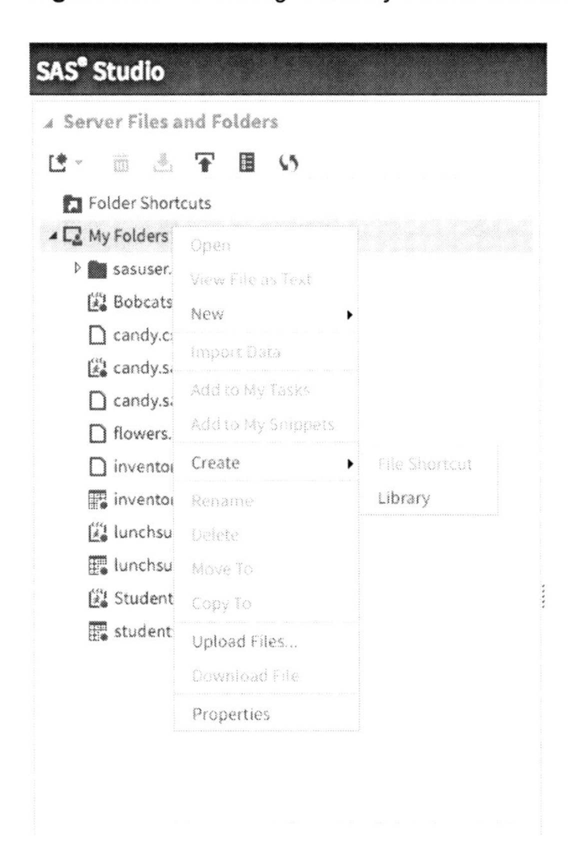

Once you select **Library**, a dialog box appears. You can name your library. (The library's name is its libref.) A libref must be 8 characters or fewer, begin with a letter or an underscore, and cannot contain blank spaces. See Figure 2.16 on page 40 for an example of naming the myfolders library. Once you have named the library (and subsequently created the permanent libref), select the check box to confirm you want the library to be re-created each time you start a SAS Studio session. This asks SAS to re-establish this library connection each time a SAS session is started.

Figure 2.16 *Naming Your Library*

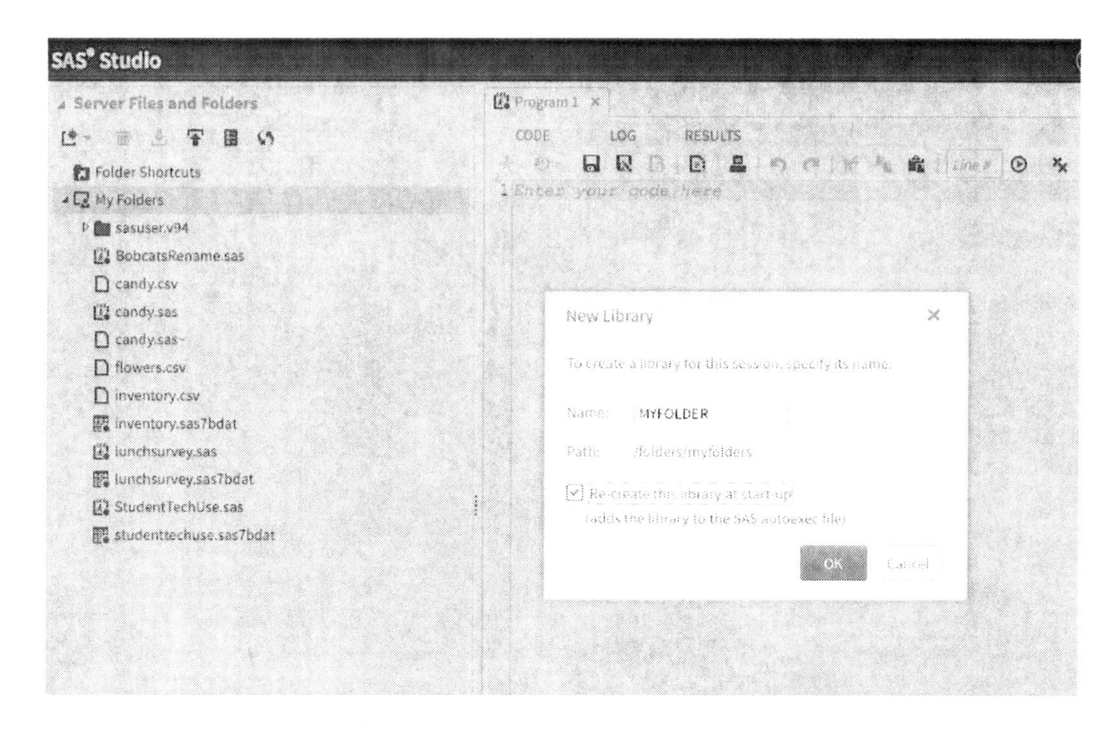

For an informative YouTube about accessing data in SAS libraries, go to http://bit.ly/1WDfl4b.

With your library established, you can bring your data into SAS! In this book, we explore several ways to import data into SAS Studio. You can import it as raw data files, select the **Import Data** menu option, or use a snippet.

Creating SAS Data Sets from Raw Data Files

Bringing in raw data using the DATA step is versatile. It enables the user to input almost any type of raw data file. To create a SAS data set from raw data, you must do the following:

1. Create and name a SAS data set (with a DATA statement).
2. Identify the location of raw data to read (with an INFILE statement).
3. Describe how to read the data fields from the raw data (with an INPUT statement).

Programming 1 and 2;

Figure 2.17 *Creating a SAS Data Set*

Raw Data File

```
         1    1    2
1---5----0----5----0
43912/11/00LAX 20137
92112/11/00DFW 20131
11412/12/00LAX 15170
```

DATA Step

```
data SAS-data-set-name;
   infile 'raw-data-filename';
   input input-specifications;
run;
```

The DATA statement tells SAS what data set you are creating. The INFILE statement tells SAS what file to read and where it is. The INPUT statement defines the variable names, variable types (character or numeric), and the physical locations for each field. There are three main types of data files when importing raw data:

- Column data
- Formatted data
- Delimited data

Figure 2.18 *Reading Data*

Column and formatted input are similar in that the values are in fixed fields.

For example, the date always starts in column 4.

```
         1    1    2
1---5----0----5----0

43912/11/00LAX 20137
92112/11/00DFW 20131
11412/12/00LAX 15170
98212/12/00dfw  5 85
43912/13/00LAX 14196
```

Delimited input, also known as **list input**, indicates that values are separated by a delimiter. The values are not in fixed fields. A comma-separated file (CSV) is an example of a delimited file.

```
         1    1    2
1---5----0----5----0---

439,12/11/00,LAX,20,137
921,12/11/00,DFW,20,131
114,12/12/00,LAX,15,170
982,12/12/00,dfw,5,85
439,12/13/00,LAX,14,196
```

In the following SAS program, the data being imported into SAS is the flowers.csv file. The CSV file is not a SAS data file, but it is a raw data file.

The DATA step specifies to create a table named **flowers** in the **myfolder** library. The INFILE statement establishes where the CSV file is located and that the delimiter is a comma. The INPUT statement specifies that there will be three variables—two character and one numeric. The delimiter is identified as **dlm** and is followed with the delimiter enclosed in quotation marks. When a raw data file is a delimited file, character variables are identified with a colon, a dollar sign, and the length of the variable.

```
DATA myfolder.flowers;
   infile '/folders/myfolders/flowers.csv' dlm = ",";
   input CustomerID : $4. Flower : $10. Quantity;
run;
```

The SAS code creates a SAS data set called **flowers** in the library **myfolder** with the character variables **CustomerID** and **Flower** and the numeric variable **Quantity**. SAS creates the data set from the raw data file flowers.csv, which was saved in **myfolder** on your computer. This example uses a CSV file, but raw data can come in many forms with many extensions, including DAT or TXT. After you run your DATA step, you can write other PROC steps to process the SAS data set.

Reading in a raw data file can also be done in SAS University Edition using point-and-click options.

Each software application has its own format for data files. Although this is good for software developers, it can be hard for software users who need to work with a variety of data sets. In particular, if you want to analyze third-party data, and the data comes in a format that is not

SAS, you have a problem. But, you can import the data into SAS by converting it. There are several converting options in SAS, including:

- Using the Import Data menu option.
- Creating a snippet.

Both of these options use PROC IMPORT and can import data easily and successfully.

Converting and Importing Raw Data into SAS Using the Import Data Menu Option

SAS Studio has a new icon that enables you to create things. You can create a new SAS program, import data, or build a query. To import data, follow these steps.

First, save your raw data file (for example, an Excel file) into My Folders on your computer.

Second, in SAS Studio, click under **Server Files and Folders** in the Navigation pane.

Third, select **Import Data**. An Import Data area appears.

Figure 2.19 *Selecting Import Data*

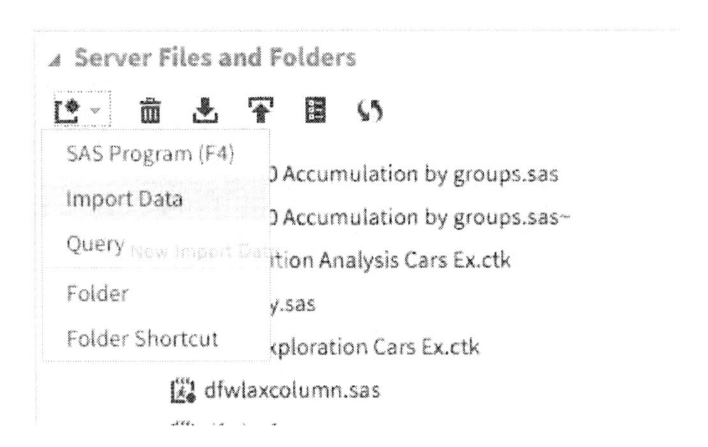

Next, select, drag, and drop the file into the Import Data area

Figure 2.20 *Selecting, Dragging, and Dropping a File*

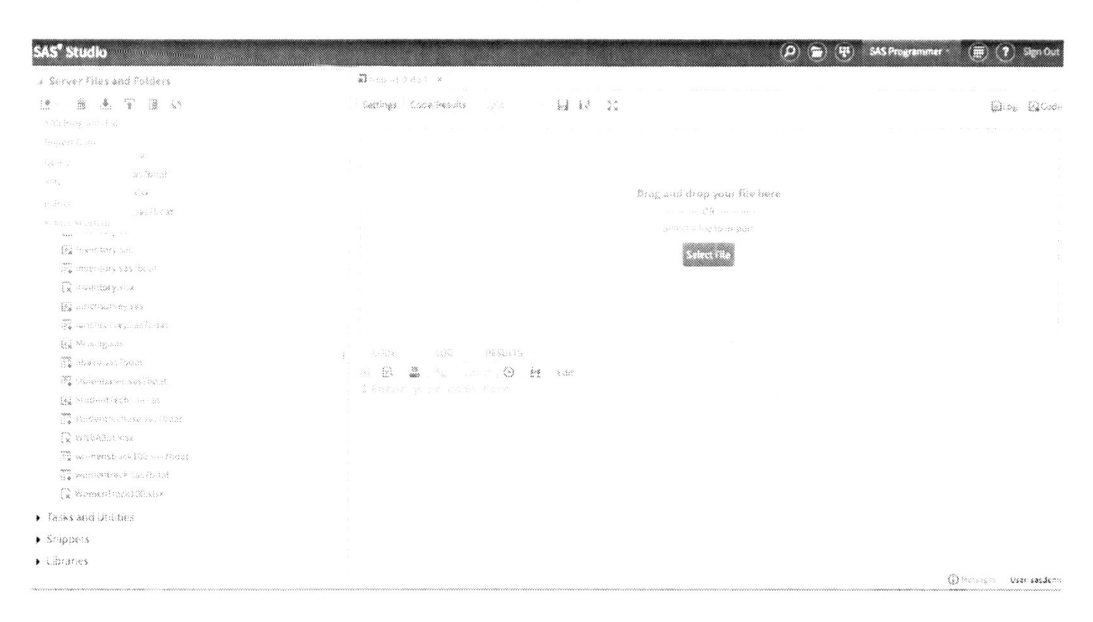

In the window, specify a name for your file and the library where you would like to save it.

Figure 2.21 *Specify Name for File*

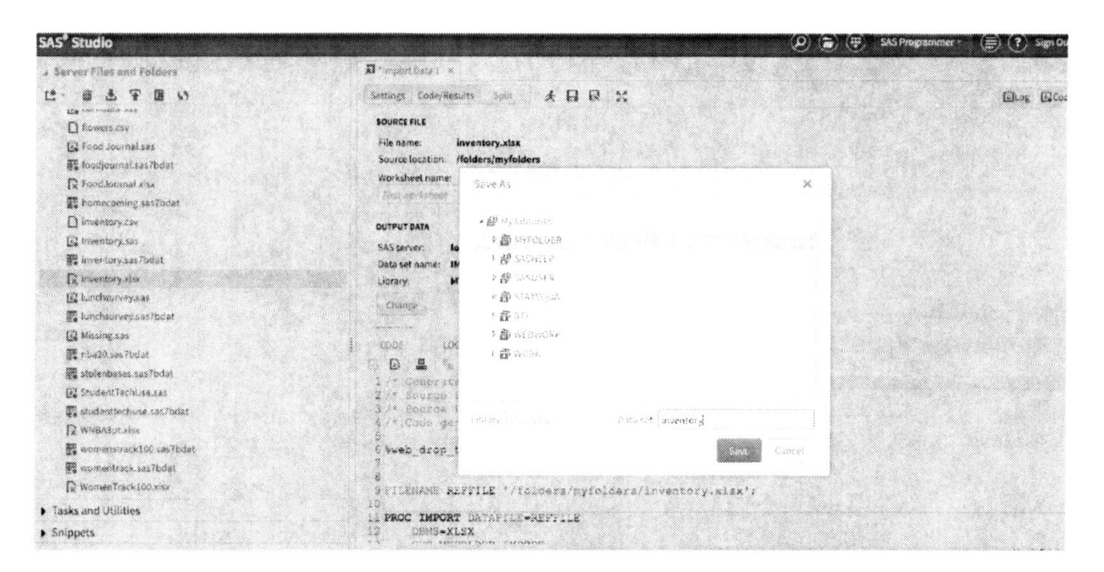

Click the running man icon to run the code.

Once you run the program, your SAS data set is created.

Converting Raw Data into SAS Data Sets Using Snippets

SAS Studio contains an Import CSV File snippet and an Import XLSX File snippet to import your data.

To import a raw data file (such as a CSV file), use the Import CSV File snippet in SAS Studio. This snippet uses PROC IMPORT. Change the values of the DATAFILE= and DBMS= options to customize code to import other types of raw data files.

Figure 2.22 *Import CSV File Snippet*

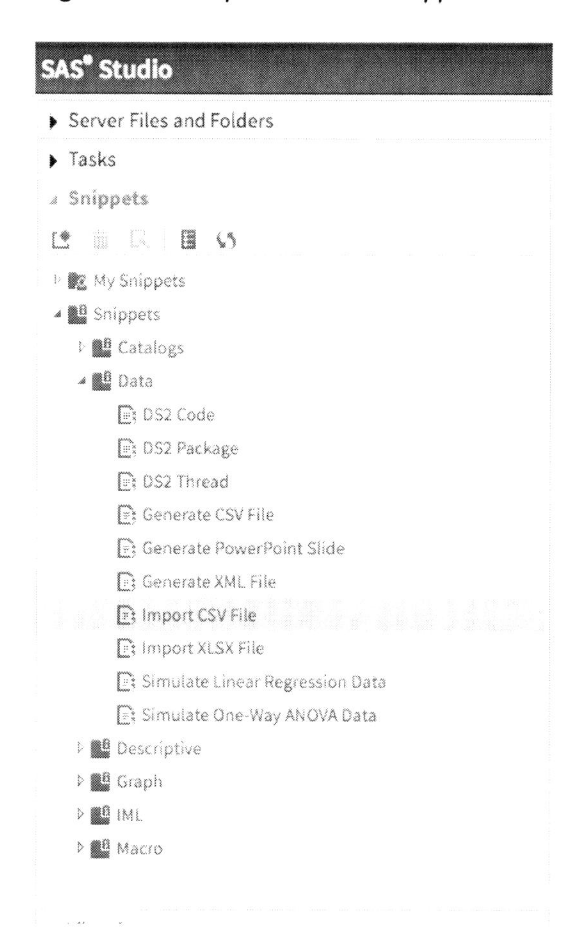

If you have data that is in the Excel format, you can upload the Excel file directly into SAS using the Import XLSX File snippet.

1. In SAS Studio, under **Tasks**, expand **Snippets**.
2. Select **Snippets > Data > Import XLSX File**. Code is added to the CODE tab. Customize the filenames in the code.

Figure 2.23 *Import XLSX File Code*

```
 1 /** Import an XLSX file.   **/
 2
 3 PROC IMPORT DATAFILE="<Your XLSX File>"
 4             OUT=WORK.MYEXCEL
 5             DBMS=XLSX
 6             REPLACE;
 7 RUN;
 8
 9 /** Print the results. **/
10
11 PROC PRINT DATA=WORK.MYEXCEL; RUN;
```

Line 3: In the PROC IMPORT statement, specify the name of your Excel file. This file must be saved in one of your folders. If your Excel file is saved in My Folders, an example filename is /folders/myfolders/inventory.xlsx. If you have created folder shortcuts, an example filename using a shortcut is /folders/myshortcuts/folder1/inventory.xlsx.

Note: You must include the forward slash at the beginning of the directory path. If you omit this forward slash, you get a “Physical file does not exist” error in the log.

Line 4: Use the OUT statement to specify where to save the SAS table. Use the format *Library.Filename*. In this example, the SAS table is a file called MYEXCEL and is saved to the Work library, which is the temporary library. If you want this data to persist, specify a different library.

Line 5: The DBMS statement specifies that you are importing an Excel file. By default, the XLSX file extension is used.

Note: If you are importing data from an older version of Microsoft Excel, change XLSX to XLS.

Line 6: The REPLACE statement specifies to overwrite an existing file with the same name.

Line 7: The RUN statement ends the PROC IMPORT step.

You can use PROC PRINT to view the new table in SAS Studio.

Analyze the Data

Now that you have your data in SAS, what should you do with it? Let’s review what we have already discussed. You have a project that you want to answer some important questions, you have identified a data set, the data has been imported into SAS, and, as part of the project planning process, you have considered the applicable code that could help get answers.

A few reminders:

- PROC PRINT displays the data in the Output window.
- PROC MEANS calculates summary statistics that include average, minimum, and maximum.
- PROC FREQ produces frequency counts and crosstabulation tables.

Understanding these procedures gives you the tools to begin to dive into your data.

Presentation Techniques

As you get ready to present your results, here are a few tips on how to impress your audience:

1. **Use visual aids:** Using pictures in your presentation instead of words can double the chances of meeting your objectives. Take advantage of the graphic capabilities in SAS and create visuals that showcase your analytical work.
2. **Keep it short and sweet:** There is an old adage that says, "No one ever complained of a presentation being too short." Nothing kills a presentation more than going on too long. Identify key points and focus on them, not every part of the process.
3. **Use the rule of three:** A simple technique to apply is that people tend to remember only three things. Decide on these three things and structure your presentation around them. Use a maximum of three points per slide.
4. **Tell a story:** A presentation is a type of theater. Tell stories and anecdotes to help illustrate points. It makes your presentation more effective and memorable.
5. **Rehearse:** Practice makes perfect. Many experts say that rehearsal is the biggest single thing that you can do to improve your performance. Perform your presentation out loud at least four times.

Quick Tips

1. Just like following a recipe for a casserole, creating a plan for a programming project is non-negotiable. Starting with good ingredients or with a good plan creates an ease of execution because you know where to begin and end.
2. There are eight phases of project planning.
3. Using PROC IMPORT enables you to import data into SAS.
4. PROC PRINT displays the data in the Output window. PROC MEANS calculates summary statistics. PROC FREQ finds the frequency of variables.
5. Tell a story with your data!

3

Basic Analytics

Analytics and Statistics–The Essentials

Grits (dried, ground corn) can be used as a base for many different recipes. It is considered a staple of Southern cooking. Most people either love or hate grits—not many fall in between!

Figure 3.1 *Cooking Grits*

Analytics and statistics are staples of the math discipline, and they are often met with the same fate of either being loved or hated. However, math is essential. We use math and statistics everyday! You use math at the grocery store; when you manage your money; and when you cook dinner, drive, walk, or choose what to watch on TV. We are all analytical people, and we often analyze things without even realizing it.

Vivian Howard, author of *A Chef's Life*, states that she has turned many grit haters into grit lovers by changing the way they perceive grits. I think the same thing can happen here as we explore the world of basic analytics.

Figure 3.2 *Delicious Southern Grits*

What Are Analytics and Statistics?

Statistics play an important role in analytics—one cannot exist without the other. *Analytics* is extracting information out of data, and *statistics* is the study of the collection, analysis, interpretation, presentation, and organization of data. We generate statistics from data, and we then use statistics to create analysis.

Organizing Data

I am from the South, and as many Southerners can attest, we think our mama's and our grandma's cooking are the best! My grandmother, Mama D., handwrote all of her recipes on small pieces of paper, and my mom handed them down to me a few years ago. The recipes were in one big envelope with no organization at all! I had to take time to go through the recipes, organize them into categories, and put them into a book for display. This was a part of my history, and I not only wanted to use the recipes, but I wanted to showcase them as well.

Just as you would organize ingredients when you are getting ready to cook a recipe, you review your collected data and establish a context for it.

Figure 3.3 *Organizing Data*

The following recipes are from a book that I created. They are organized by desserts and cakes. My Papa loved the Hummingbird cake. My children like the Gum Drop cake, but without the walnuts. I manipulate the Gum Drop cake recipe so that it works for them!

Figure 3.4 *Recipes Organized by Category*

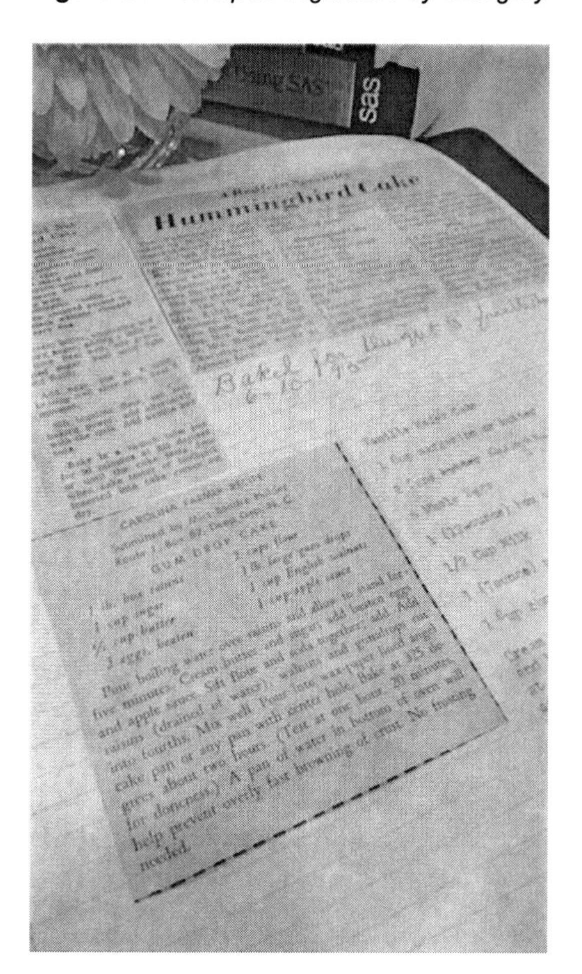

Organizing and displaying recipes are not any different from understanding and staging your data. I created categories, much like variables, to understand and sort what was in the envelope. When you receive data, you want to understand the content. Understanding data is as simple as reading the recipe above. If you know your variables and understand the context of the questions asked, you can comprehend the observations. From the Gum Drop cake recipe, you know you have eight ingredients. Once we can decipher the contents, we are ready to formulate questions to drive the analysis.

Data is the foundation for all investigations, no matter the field or discipline. Data consists of variables and observations. Understanding the data enables you to begin analyzing with efficiency.

There are two types of data—quantitative and qualitative.

- *Quantitative* data is the result of using an instrument (survey, scientific measure, etc.) to measure something (test score, weight, sales, age, etc.).

- *Qualitative* data is also referred to as frequency data. This data looks at observations that are grouped based on a common property, and then the number of members of the group are recorded (for example, males and females, vehicle type, product type, etc.).

Tasks in SAS University Edition

The Navigation Pane

Before we dive into the different analytical tools, let's look at the Navigation pane in SAS Studio. To run analytics, look under **Tasks and Utilities**.

Figure 3.5 *Tasks and Utilities Menu*

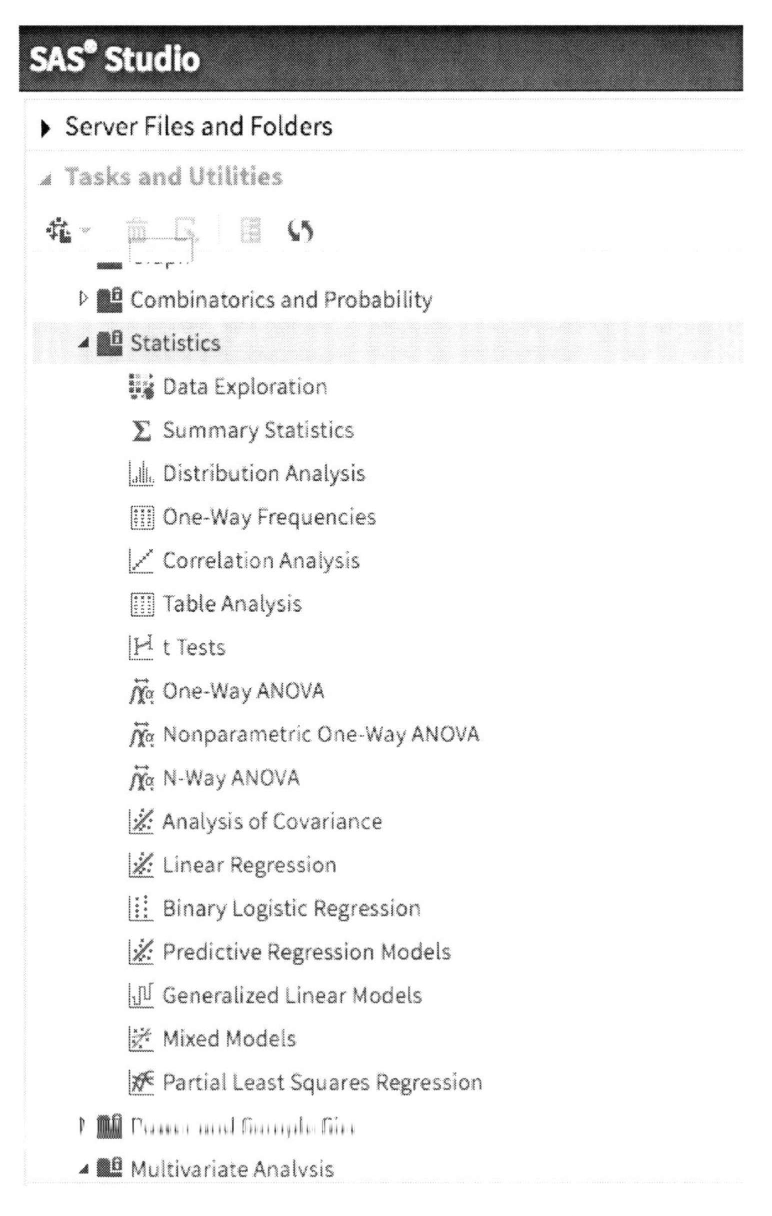

Under **Tasks and Utilities,** a drop-down list for **Statistics** offers analytical tools (such as **Data Exploration**, **Summary Statistics**, and **Distribution Analysis**) that you can use to work with your data. The tasks are point-and-click interfaces that generate SAS code that reflects decisions that you make and options that you selected.

Selecting a Task

Once you select a task, you enter task settings. Settings change based on the analytical tool. You enter settings in the expanded Work area (which is shown below). By default, both the settings for the selected task and code are shown (in split screens of the Work area). In Figure 3.6 on page 56, the DATA tab and the Split view tab are selected.

Figure 3.6 *Summary Statistics under the Tasks and Utilities Menu*

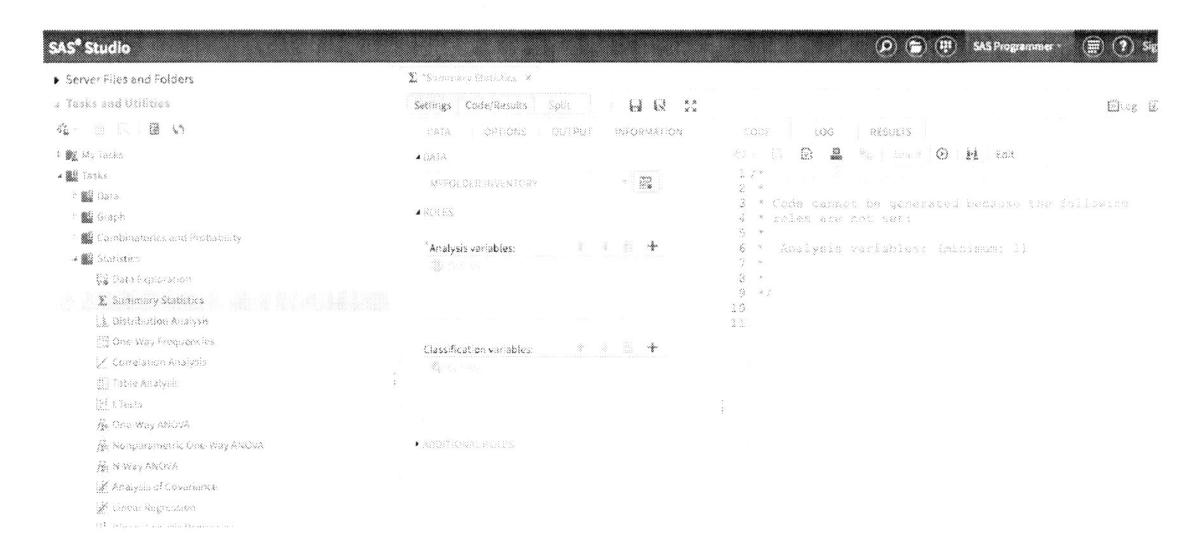

The DATA Tab

The DATA tab is where you select the data that you want to use for the task. In Chapter 2, you imported the data into SAS and it is now a SAS data set. By clicking , you can navigate and select a data set. In the data set, you can assign variables to roles. Roles vary by task, but variables include choices such as:

- Analysis variables
- Continuous variables
- Classification variables
- Dependent variables

Some roles have variable limitations, and that is shown within each task.

To assign a variable to a role, click +, and select a variable. Multiple variables can be selected by holding down the Shift and Ctrl keys while selecting. To delete a variable, select the variable in the text box, and click .

At the bottom of the DATA tab, additional roles can be selected. These vary based on the task.

While you select roles and settings on the left side of the Work area, SAS code is being generated on the right side. For each task, there are certain roles that must be populated in order for the code to be generated. When you are ready to run the code, click , and the output will appear on the OUTPUT tab. Figure 3.7 on page 57 shows the **Summary Statistics** task, specifically PROC MEANS. In the Work area, settings are on the left and generated code is on the right.

Figure 3.7 *Summary Statistics Task with Generated Code*

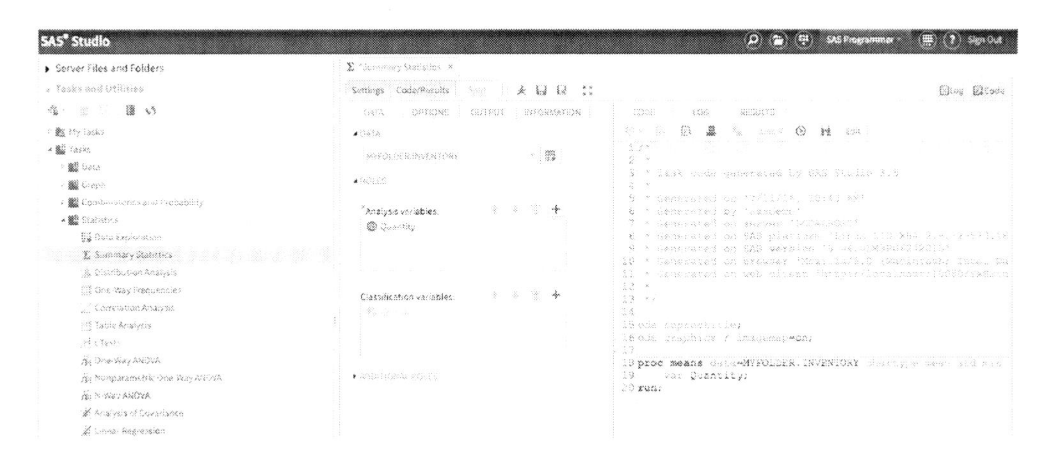

The OPTIONS Tab

You might want to change a few settings that have been set by default. On the OPTIONS tab, you can add plots or details or include or exclude certain statistics. Figure 3.8 on page 58 shows the OPTIONS tab.

Figure 3.8 *OPTIONS Tab*

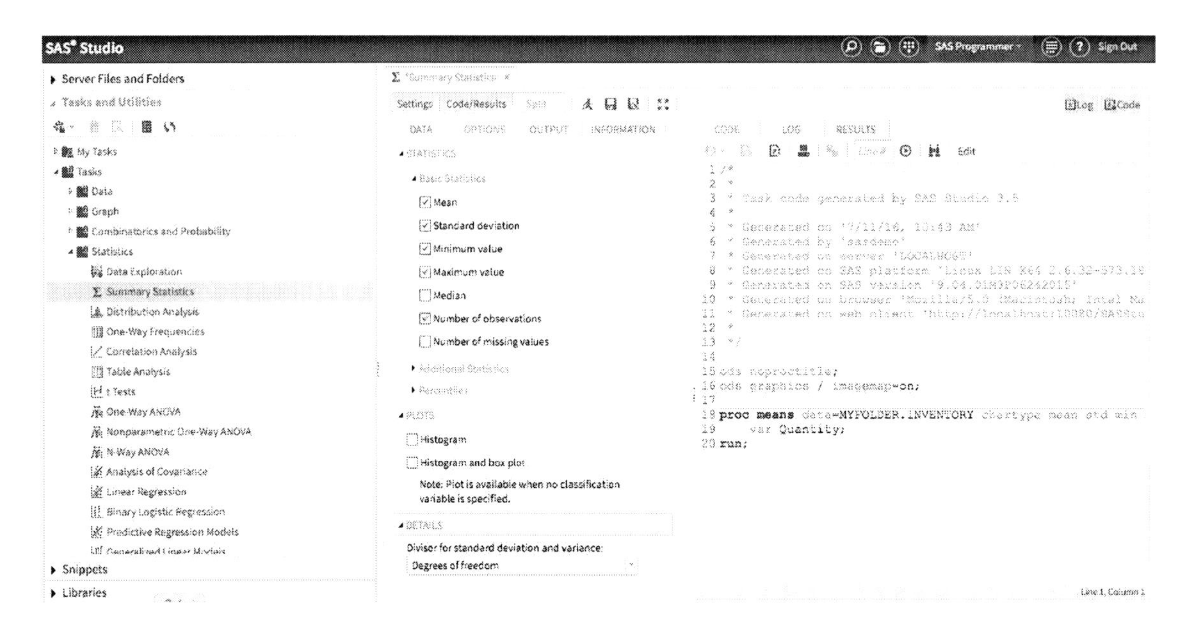

When you have completed all of your selections on the OPTIONS tab, rerun your code by clicking .

The OUTPUT Tab

Some tasks enable you to create a new data set as part of the results. By default, the data set is placed in the Work library. It is a subset of the original data set. Not all tasks include the OUTPUT tab. In “Working with Statistics” on page 59, the new data set name is identified.

Figure 3.9 OUTPUT Tab

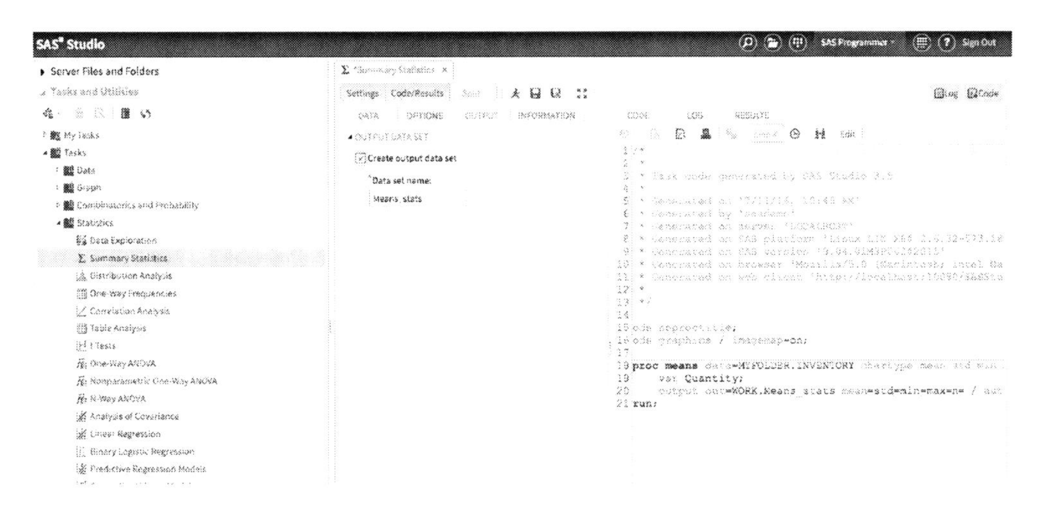

Working with Statistics

Statistics is the baseline for performing analytics. Statistics give us the vehicle to find the novelty in our data. If we overdo or overcomplicate our analysis, then the value of the analysis has been lost. Therefore, we need to understand and effectively use the resources in SAS University Edition.

Introduction to Variables and Graphs

Variables

A variable is the property of an object or event that can have different values. For example, college major is a variable that can have different values such as mathematics, computer science, or biology. Variables can be categorized based on their information. Below is a breakdown of the different categories of variables.

A numeric or quantitative variable has a value that describes a measurable quantity as a number, such as “How many?” or “How much?” A numeric variable is either discrete or continuous.

- A discrete variable has a limited number of values.

- A continuous variable can have many different values. In theory, it can have *any* value between the lowest and highest points on the measurement scale.

A character or qualitative variable has a value that describes a quality or characteristic of a data unit, such as "Which type?" or "Which category?" You will also see this called a categorical variable when using statistics. Character or qualitative variables are either mutually in one category or another or exhaustive, which includes all possible categories. A character variable has a non-numeric value. A character variable can be ordinal or nominal.

- An ordinal variable has a value that can be logically ordered or ranked. It can be ranked as higher or lower, but it does not necessarily establish a numeric difference between categories. (For example, an academic grade can be an A or B; a clothing size can be small, medium, or large; and an attitude can be agree or strongly agree.)
- A nominal variable cannot be organized in a logical sequence (for example, business type, eye color, brand, etc.).

Note: Some numeric variables can be character variables.

Figure 3.10 *Variables Flowchart*

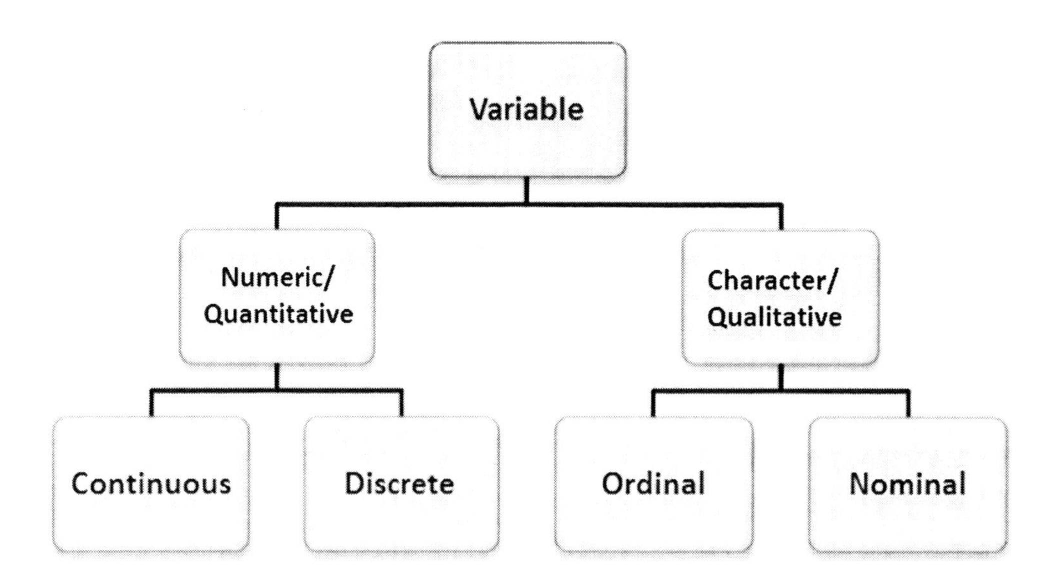

Variables can be complex in terms of category. As you evaluate the observations in your data set and you are able to establish the defined category for the data and variable, you might need to run analysis by defining the variable as independent or dependent.

- An independent variable can be manipulated, measured, or selected by the researcher as an antecedent condition to an observed behavior. In a hypothesized cause-and-effect relationship, the independent variable is the cause and the dependent variable is the outcome or effect.

- A dependent variable is not under the researcher's control. It is the variable that is observed and measured in response to the independent variable.

An independent variable is the presumed cause, whereas the dependent variable is the presumed effect. Another way of looking at it is the independent variable is the variable that is varied, and the dependent variable is the response that is measured. For example, does a person's weight (independent) have an impact on his cholesterol (dependent), sugar (dependent), iron (dependent), etc.?

Another example of independent and dependent variables is when a scientist studies the impact of a drug on blood cancer. The independent variable is the administration of the drug, and the dependent variable is the impact the drug has.

Variables drive the data and are the entities that you work with to complete your analysis.

To visually see the variables in your data set, create a graphical display.

Graphs

A graph is a visual display of data that presents frequency distributions. In a graph, the shape of the distribution can easily be seen. The visual aspect of a graph is powerful in that it showcases the data that you have analyzed. Below is a breakdown of different graphs that can be used to display data. After each definition is a visual of the graph showcasing the data set myfolder.inventory.

Under **Tasks and Utilities**, select **Graph**, and then select the graph of your choice.

Figure 3.11 *Graph Task*

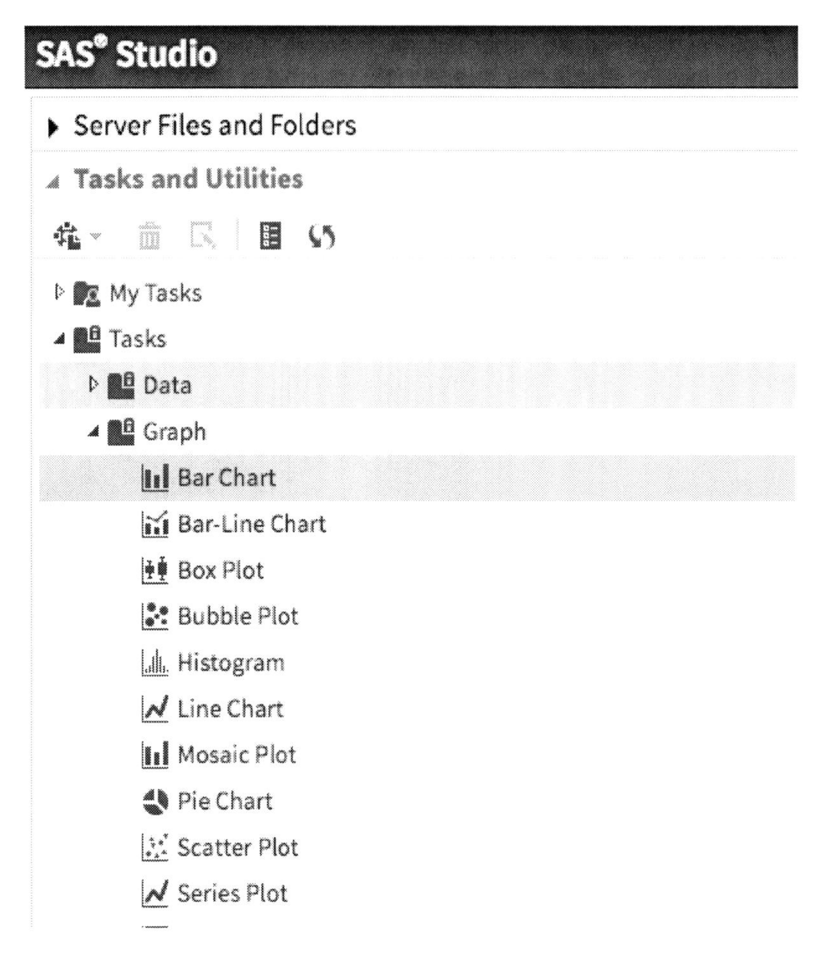

Here are explanations for four of the most popular graphs:

Bar Chart: In a bar chart, the higher the bar, the higher the frequency of occurrence. The underlying measurement scale is discrete (nominal- or ordinal-scaled data), not continuous.

Figure 3.12 *Bar Chart Coding*

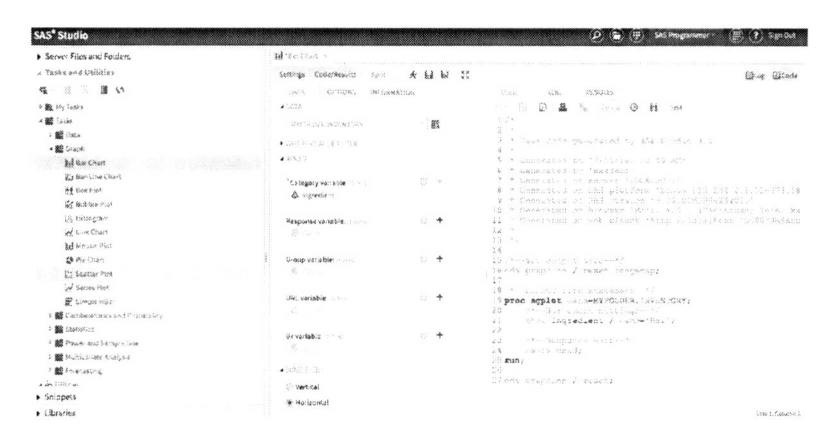

The bar chart represents the character or qualitative variable, Ingredient. The variable is nominal because it is not in an ordered sequence.

Figure 3.13 *Bar Chart Example*

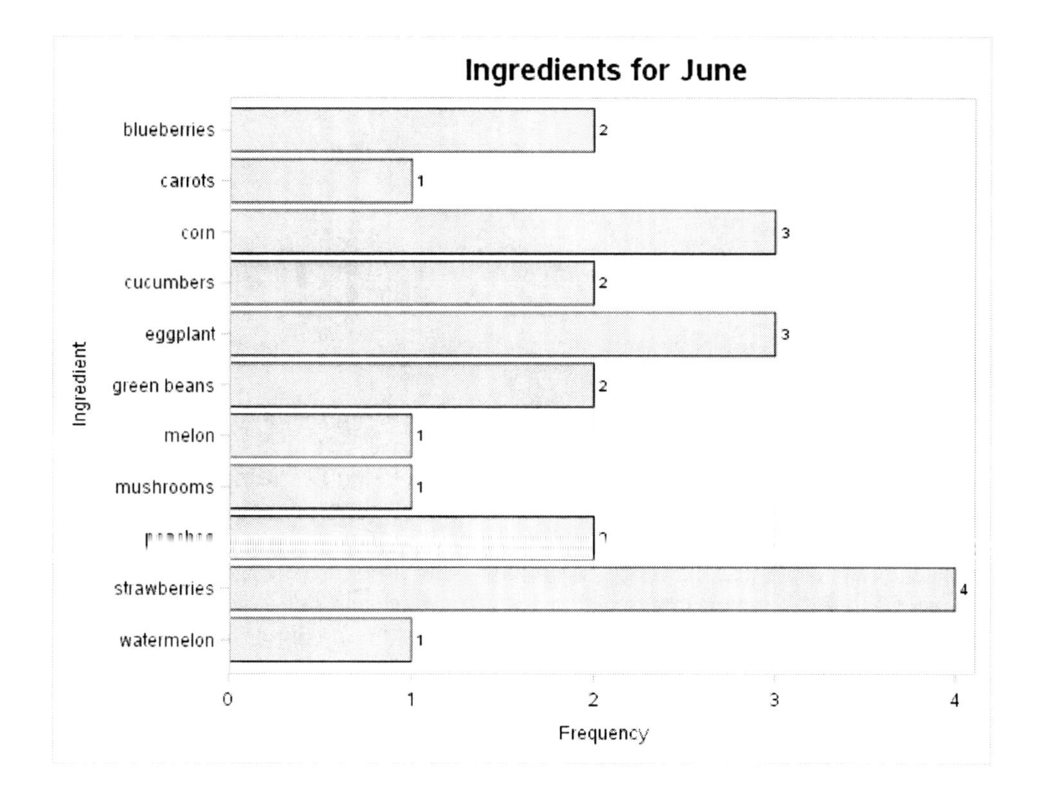

Histogram: A histogram is a form of bar chart used with interval- or ratio-scaled data. Unlike the bar chart, bars in a histogram touch each other, with the width of the bars defined by the upper and lower limits of each interval. The measurement scale is continuous, so the lower limit of any interval touches the upper limit of the previous interval.

In Figure 3.14 on page 64, the variable is **Quantity**, which is a numeric or quantitative variable and continuous. The quantity can change from month to month, which is the characteristic that makes it a continuous numeric variable.

Figure 3.14 *Histogram Example*

Box Plot: A box plot is a graphical representation of dispersions and extreme scores. In the following box plot, minimum, maximum, and quartile scores are represented in the form of a box with whiskers. The box includes the range of scores falling into the middle 50% of the distribution (interquartile range (IQR) = 75th percentile - 25th percentile). The whiskers extend to the last observation value that is within 1.5*IQR of Q1 and Q3. This might be the maximum and minimum scores in the distribution, but it depends on your data.

In Figure 3.15 on page 65, the variable is Quantity, which is a numeric or quantitative variable and continuous. The quantity can change from month to month, which is the characteristic that makes it a continuous numeric variable.

Figure 3.15 *Box Plot Example*

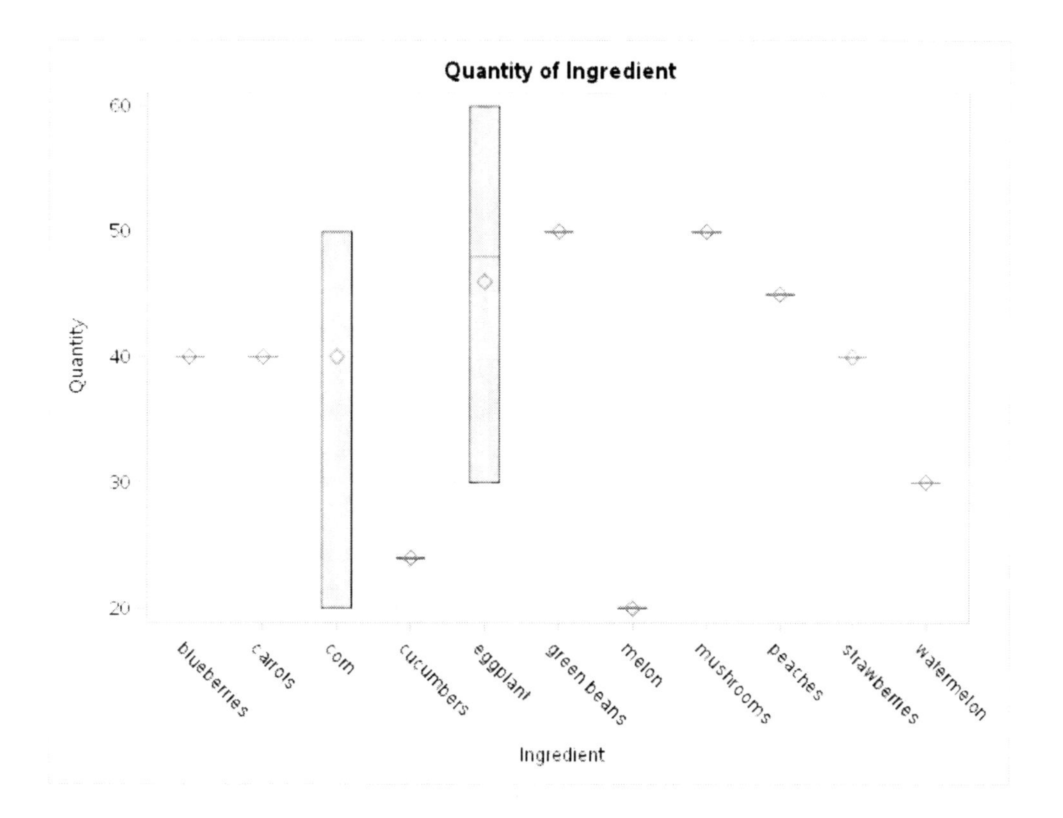

Scatterplot: A scatterplot presents information from a bivariate distribution. ("Bi" means two, so a bivariate distribution looks at two variables.) In a scatterplot, each subject in an experimental study is represented by a single point in a two-dimensional space. The underlying scale of measurement for both variables is continuous. This is one of the most useful techniques for gaining insight into the relationship between two variables.

In Figure 3.16 on page 66, the variables are Ingredient and Quantity. The scatterplot can display either a numeric or quantitative variable or a character or qualitative variable.

Figure 3.16 *Scatterplot Example*

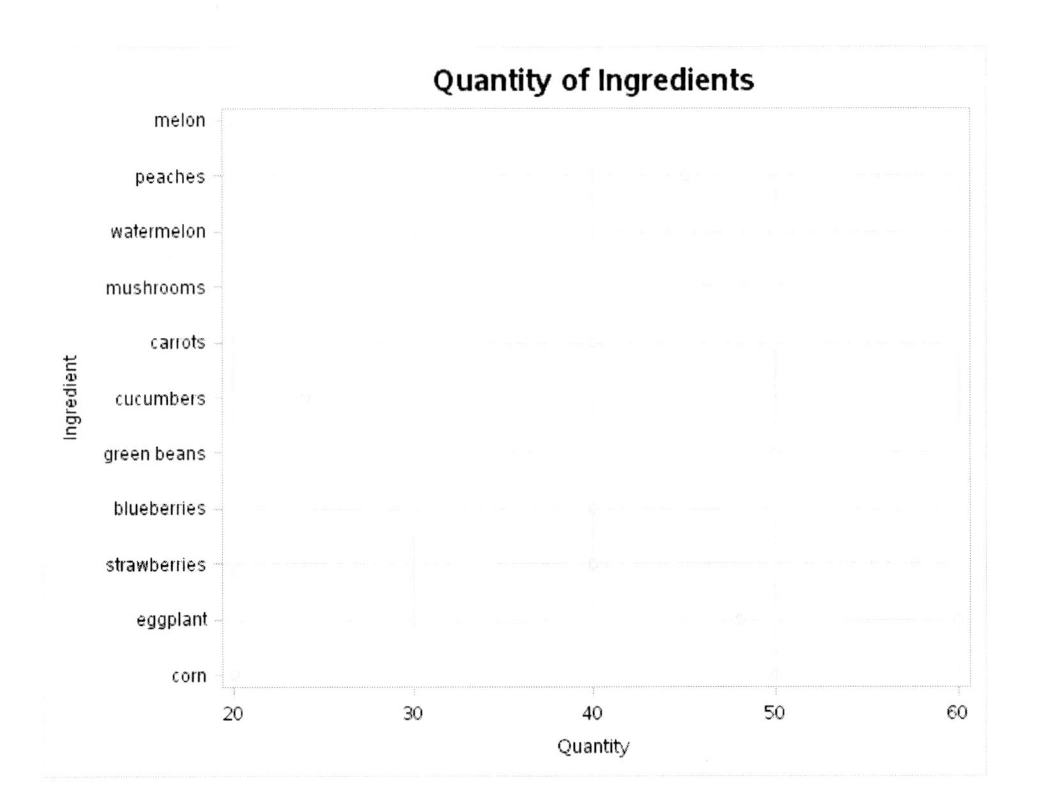

Summary Statistics

Introduction

In Chapter 2, we discussed several procedures and why you might use them. In this section, we delve into more detail, touching on PROC UNIVARIATE, PROC MEANS, and PROC FREQ. The myfolder.inventory data that we have been using throughout the book is used in these procedures as well.

PROC UNIVARIATE

Introduction

The following PROC UNIVARIATE returns an extensive summary of the data. It can also generate graphs.

```
PROC UNIVARIATE data=myfolder.inventory;
    Title "Summary of Data";
run;
```

In the code, PROC UNIVARIATE is run and analyzes all of the numeric variables in the data set. When it is run, five tables get populated.

1. Moments table
2. Basic Statistical Measures table
3. Tests for Location table
4. Quantiles table
5. Extreme Observations table

You can run PROC UNIVARIATE by expanding **Statistics** under **Tasks and Utilities**, selecting **Distribution Analysis**, and not writing any code. The results are the same.

Figure 3.17 PROC UNIVARIATE via the Tasks and Utilities Menu

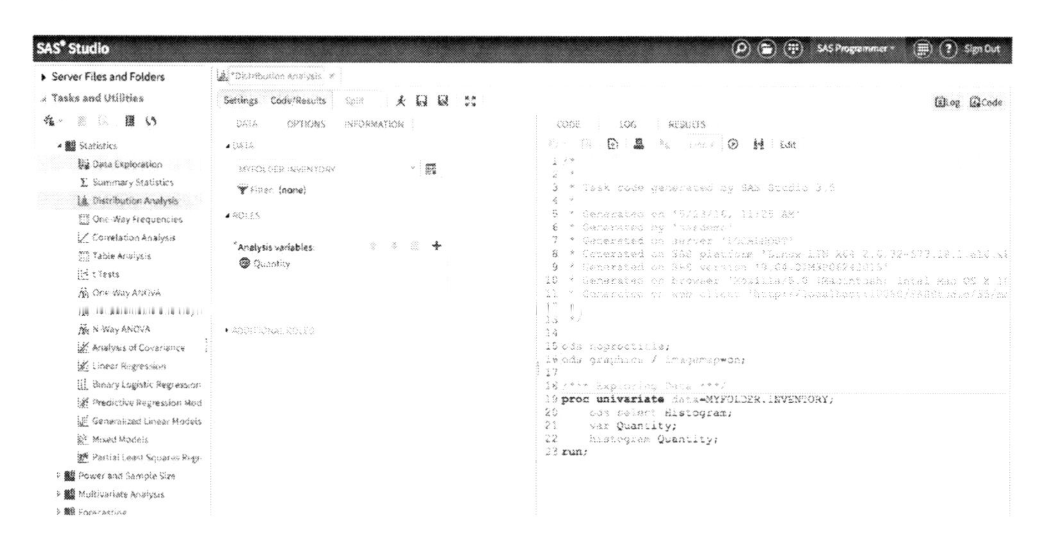

Figure 3.18 *PROC UNIVARIATE Results*

Summary of Data

The UNIVARIATE Procedure
Variable: Quantity

Moments			
N	22	Sum Weights	22
Mean	39.8181818	Sum Observations	876
Std Deviation	10.9659054	Variance	120.251082
Skewness	-0.4390342	Kurtosis	-0.5090795
Uncorrected SS	37406	Corrected SS	2525.27273
Coeff Variation	27.5399452	Std Error Mean	2.3379389

Basic Statistical Measures			
Location		Variability	
Mean	39.81818	Std Deviation	10.96591
Median	40.00000	Variance	120.25108
Mode	40.00000	Range	40.00000
		Interquartile Range	20.00000

<table>
<tr><th colspan="5">Tests for Location: Mu0=0</th></tr>
<tr><th>Test</th><th colspan="2">Statistic</th><th colspan="2">p Value</th></tr>
<tr><td>Student's t</td><td>t</td><td>17.03132</td><td>Pr > |t|</td><td><.0001</td></tr>
<tr><td>Sign</td><td>M</td><td>11</td><td>Pr >= |M|</td><td><.0001</td></tr>
<tr><td>Signed Rank</td><td>S</td><td>126.5</td><td>Pr >= |S|</td><td><.0001</td></tr>
</table>

Figure 3.19 *PROC UNIVARIATE Results (continued)*

Quantiles (Definition 5)	
Level	Quantile
100% Max	60
99%	60
95%	50
90%	50
75% Q3	50
50% Median	40
25% Q1	30
10%	24
5%	20
1%	20
0% Min	20

Extreme Observations			
Lowest		Highest	
Value	Obs	Value	Obs
20	15	50	5
20	14	50	7
24	21	50	11
24	6	50	20
30	16	60	2

Let's look at each of the tables.

The Moments Table

The Moments table includes the following information:

- The mean, which is the average of the data.
- The standard deviation, which is the square root of the variance. The standard deviation is the most common measure used to describe dispersion (how stretched or squeezed the sample is) or variability around the mean. When all the values are close to the mean, the standard deviation is small. When the values are scattered widely from the mean, the standard deviation is large. Like variance, standard deviation is a measure of dispersion around the mean, but it is easier to interpret because the measurements are in the same units as the data.

- The skewness is the measure of the tendency for the distribution of values to be more spread out on one side than on the other side. Positive skewness indicates that values to the right of the mean are more spread out than values to the left. Negative skewness indicates that values to the left of the mean are more spread out than values to the right.
- The coefficient of variation is also known as the relative standard deviation (RSD). It is a standardized measure of dispersion of a distribution. Coefficient of variation is calculated as the standard deviation divided by the mean and multiplied by 100.
- The sum of weights is the same as N because each observation is given equal weight. N is the number of observations in the data set.
- The sum of observations is the sum of the values for all rows.
- The variance is the square of the standard deviation. The variance is never less than 0. It measures how far a set of numbers is spread out. A variance of 0 indicates that all values are the same. A small variance indicates that the data points tend to be very close to the mean and to each other. A high variance indicates that the data points are very spread out around the mean and from each other.
- The kurtosis is the measure of the shape of the distribution of values. Kurtosis measures how peaked or flat the shape of the distribution of values is. For example, large values for kurtosis indicate that the distribution has heavy tails or is flat. This means that the data contains some values that are very far from the mean.
- The standard error of the mean is calculated the same way as standard deviation, but it is divided by the square root of the sample size. It is used to calculate the confidence interval. It is the variability of the mean due to sampling variability.

The Basic Statistical Measures Table

The Basic Statistical Measures table has two columns. The Location column shows statistics that summarize the center of the distribution of the values. The Variability column shows statistics that summarize the spread of the distribution of values. Much of the same information is repeated from the Moments table.

The Basic Statistical Measures table includes the following information:

- The mean, which is the average of the data.
- The median, which is the 50th percentile. Half of the data values are above the median and the other half are below.
- The mode, which is the data value with the largest associated number of observations.
- The standard deviation, which is the square root of the variance. The standard deviation is the most common measure used to describe dispersion (how stretched or squeezed the sample is) or variability around the mean. When all the values are close to the mean, the standard deviation is small. When the values are scattered widely from the mean, the standard deviation is large. Like variance, standard deviation is a measure of dispersion about the mean, but it is easier to interpret because the measurements are in the same units as the data.

- The variance is the square of the standard deviation. The variance is never less than 0. It measures how far a set of numbers is spread out. A variance of 0 indicates that all values are the same. A small variance indicates that the data points tend to be very close to the mean and to each other. A high variance indicates that the data points are very spread out around the mean and from each other.
- The range is the difference between the highest and lowest data values.
- The interquartile range (IQR) is the difference between the 25th percentile and the 75th percentile.

The Tests for Location Table

The Tests for Location table displays a number of statistical tests that determine whether a variable's measure of central location is significantly different from its theoretical value (mu). The default value for mu is 0.

The Quantiles Table

The Quantiles table gives more detail about the distribution of data values. The 0th percentile is the lowest value and is labeled Min. The 100th percentile is the highest value and is labeled Max. The difference between the two values is the range. Similarly, the 25th percentile is the first quartile. It is greater than or equal to 25% of the values. The 75% percentile is the third quartile. It is greater than or equal to 75% of the values. The difference between these two values is the interquartile range (IQR).

The Extreme Observations Table

The Extreme Observations table lists the five lowest and five highest observations in the lowest and highest columns. The lowest value is the minimum and the highest value is the maximum.

Why use PROC UNIVARIATE? This procedure is useful if you want an extensive summary of the data. It provides a good view of how the data is falling and what are the data's highs and lows.

However, if you need a brief summary of the data, PROC MEANS is the way to go.

PROC MEANS

PROC MEANS runs on numeric variables and provides several statistics, such as the average, maximum, and minimum. PROC UNIVARIATE provides in-depth analysis for the distribution, whereas PROC MEANS provides a more concise overview.

```
PROC MEANS data=myfolder.inventory;
    Var quantity;
    /* the variable quantity will be run */
Run;

PROC MEANS data=myfolder.inventory;
```

```
Run; /* if no variable is designated, the procedure will be run for all numeric variables */
```

PROC MEANS output is much shorter than PROC UNIVARIATE. The default results show the number of observations, mean, standard deviation, minimum, and maximum.

You can run PROC MEANS by expanding **Statistics** under **Tasks and Utilities**, selecting **Summary Statistics**, and not writing any code. The results are the same.

Figure 3.20 *PROC MEANS via the Tasks and Utilities Menu*

Figure 3.21 *PROC MEANS Results*

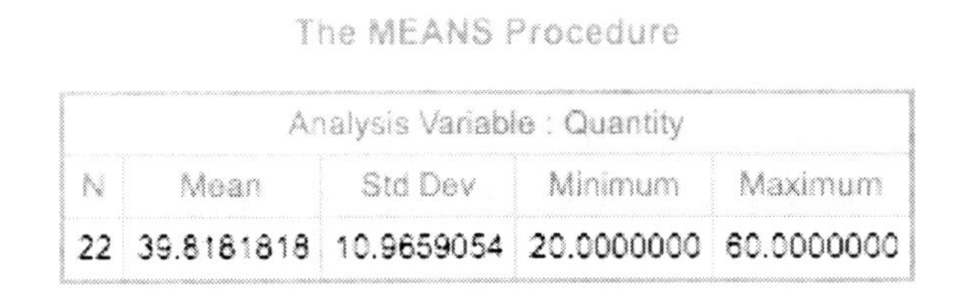

The MEANS Procedure

Analysis Variable : Quantity				
N	Mean	Std Dev	Minimum	Maximum
22	39.8181818	10.9659054	20.0000000	60.0000000

PROC MEANS output has a line that includes basic summaries. However, you can specify other options in the PROC MEANS statement, such as N, NMISS, RANGE, QRANGE, MIN, and MAX.

```
PROC MEANS data=myfolder.inventory n nmiss range qrange mean median min max;
    var quantity;
run;
```

Figure 3.22 *PROC MEANS Results with Additional Options*

The MEANS Procedure

Analysis Variable : Quantity							
N	N Miss	Range	Quartile Range	Mean	Median	Minimum	Maximum
22	0	40.0000000	20.0000000	39.8181818	40.0000000	20.0000000	60.0000000

- N is the number of observations with nonmissing values.
- N Miss is the number of missing values.
- Range is the difference between the highest and lowest data values.
- Quartile Range is the difference between the 25th percentile and the 75th percentile.
- Minimum is the minimum number.
- Maximum is the maximum number.

TIP Remember, the VAR statement tells SAS which variables to analyze. If you omit the VAR statement, SAS runs the procedure on all numeric variables in the data set.

PROC MEANS is very helpful in providing quick and meaningful statistics that can guide your project.

Confidence Interval

You might need to find the confidence interval and the standard error for your data. These two statistics help determine how well your sample mean estimates the mean of the population in which you took your sample. A confidence interval provides an estimate interval, which gives you a measure of precision or confidence around your point of estimate. This means that there is a range of values around the point of estimate within which the true value is expected to fall. Standard error refers to the standard deviation of various sample statistics, such as the mean or median. Consider an example:

A study was conducted to look at the health of a college population at a local university. Through surveys, the students were asked a variety of questions, including what their BMI (body mass index) was. Based on the data collected, the analysis showed that 36.7% of the population surveyed was overweight, with a BMI between 25.0 and 29.9. This was calculated with a 95% confidence interval of 31.6 and 41.8, or within a 5-point interval from the mean. This percentage (36.7%) is an estimate of the true population of the university students who are overweight. Because 100% of the student population was not surveyed, we use the statistics from the sample to infer the true proportion of students who are overweight.

To calculate this statistic in SAS, include the option CLM (confidence limit for the mean):

```
PROC MEANS data=myfolder.inventory n mean clm stderr;
```

```
    var quantity;
run;
```

PROC FREQ

PROC FREQ is used mostly with character variables. However, the procedure can calculate unique values for either character or numeric variables. The TABLES statement identifies which variables you want to process and displays the data in a table format. By default, PROC FREQ outputs frequencies, percentages, cumulative frequencies, and cumulative percentages. PROC FREQ also reports any missing values from the data set. The following example involves a one-way frequency table:

```
PROC FREQ data=myfolder.inventory;
    Tables ingredient vendorid;
run;
```

You can run PROC FREQ by expanding **Statistics** under **Tasks and Utilities**, selecting **One-Way Frequencies**, and not write any code. The results are the same.

Figure 3.23 *PROC FREQ via the Tasks and Utilities Menu*

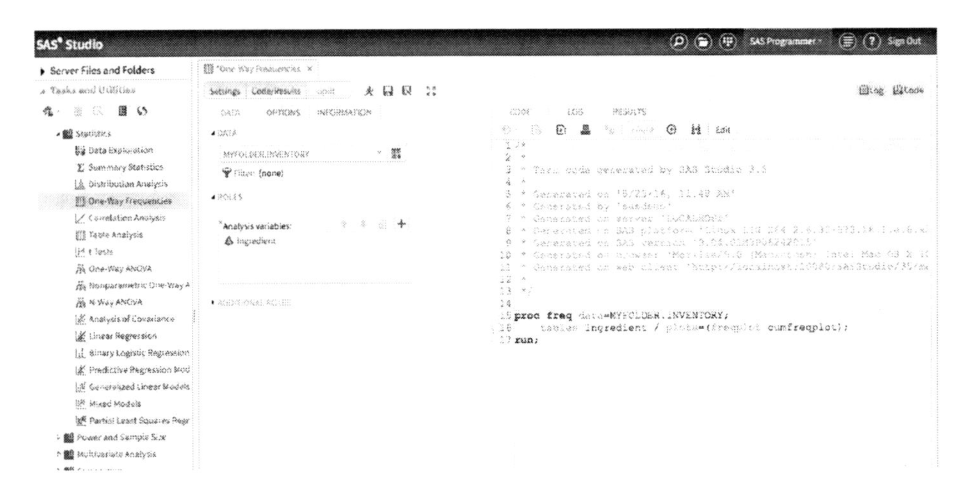

Figure 3.24 *PROC FREQ Results*

The FREQ Procedure

Ingredient	Frequency	Percent	Cumulative Frequency	Cumulative Percent
blueberries	2	9.09	2	9.09
carrots	1	4.55	3	13.64
corn	3	13.64	6	27.27
cucumbers	2	9.09	8	36.36
eggplant	3	13.64	11	50.00
green beans	2	9.09	13	59.09
melon	1	4.55	14	63.64
mushrooms	1	4.55	15	68.18
peaches	2	9.09	17	77.27
strawberries	4	18.18	21	95.45
watermelon	1	4.55	22	100.00

VendorID	Frequency	Percent	Cumulative Frequency	Cumulative Percent
188R	4	18.18	4	18.18
240W	6	27.27	10	45.45
356W	6	27.27	16	72.73
490R	2	9.09	18	81.82
501W	4	18.18	22	100.00

If you do not want to include cumulative frequencies, specify the NOCUM option in your initial PROC FREQ statement:

```
PROC FREQ data=myfolder.inventory nocum;
    Tables ingredient vendorid;
run;
```

PROC FREQ is one of the most common analytical tools because it is so versatile. You can run a crosstabulation table or a two-way table between two variables. The table is useful for analyzing the various values of a variable in a data set and in comparison to another variable. It provides for mutual exclusivity and a percentage breakdown of the combination of responses. The following chart is from a data set called work.foodinterest. It was a survey given to 100 people about their food interests. This data set was saved in the temporary Work library. Here are two questions from the survey:

1. Do you like casseroles?
2. Do you like grits?

The questions were labeled as the LIKECASSEROLES variable and the LIKEGRITS variable.

```
PROC FREQ data=work.foodinterest nocum;
    Tables likecasseroles * likegrits;
run;
```

Figure 3.25 *Table of Food Interests*

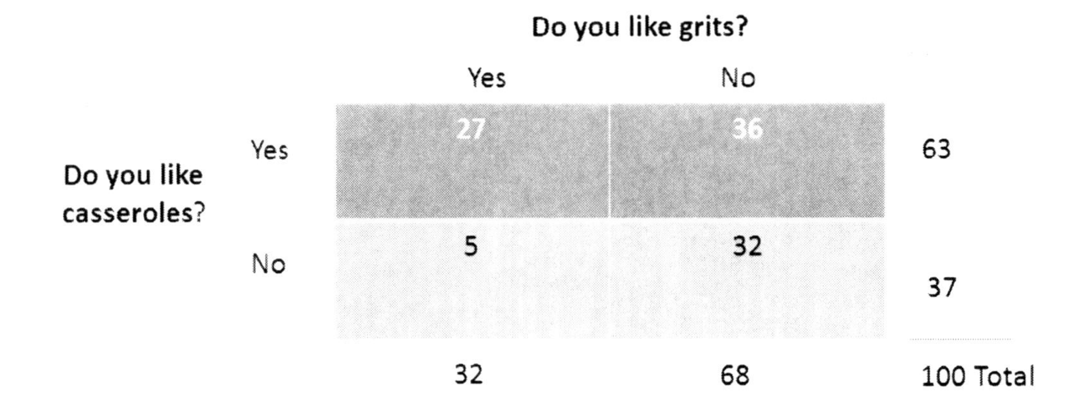

		Do you like grits?		
		Yes	No	
Do you like casseroles?	Yes	27	36	63
	No	5	32	37
		32	68	100 Total

From the two-way frequency table, the responses for the first question are listed in the rows of the table. The responses for the second question are listed in the columns. The number of participants who like both casseroles and grits is 27. The number of participants who like casseroles but not grits is 36. The number of participants who like grits but not casseroles is 5. Participants who do not like either is 32. The totals for the rows and columns should equal the total number of people surveyed.

Tests of Association

Introduction

To look for relationships in your data, you might want to run a test of association. Examples include:

- *t* test

- correlation
- regression
- analysis of variance

t Test

PROC TTEST can perform one-sample *t* tests, two-sample *t* tests, and paired *t* tests. You use a *t* test when you want to compare the means of numeric variables. PROC TTEST is a bigger step than PROC MEANS. For example, a farmer could select a few rows of his crops to test a new organic pesticide. He could test whether the true mean of the growth of the crop was above a specific level (a one-sample *t* test). Then, the farmer could compare the results with the rest of the crop that did not receive the organic pesticide (a two-sample *t* test for two independent samples). He could compare the growth of the crop before and after the pesticide (a paired *t* test).

The one-sample comparison computes a *t* test for a single mean. You list the variable in a VAR statement. SAS tests to see whether the mean is significant from the H0, which is a specified null value. The default value is 0.

With options, you can specify whether you want to change the H0 or whether you want a two-tailed *t* test (sides=2) or a one-tailed *t* test (sides=U). If you are using a significance level of .05, then the two-tailed *t* test allots half of your alpha testing to testing the statistical significance in one direction. The other half tests the significance in the other direction. The alpha, by default, is .05. In the code, the data set myfolder.inventory has sides=U, which tests whether the mean quantity purchased is more than 40 pounds.

```
PROC TTEST data=myfolder.inventory sides=U h0=40 alpha=.05;
    var Quantity;
run;
```

You can run *t* tests by expanding **Statistics** under **Tasks and Utilities**, selecting **T Tests**, and not writing any code. The results are the same.

Figure 3.26 *PROC TTEST via the Tasks and Utilities Menu*

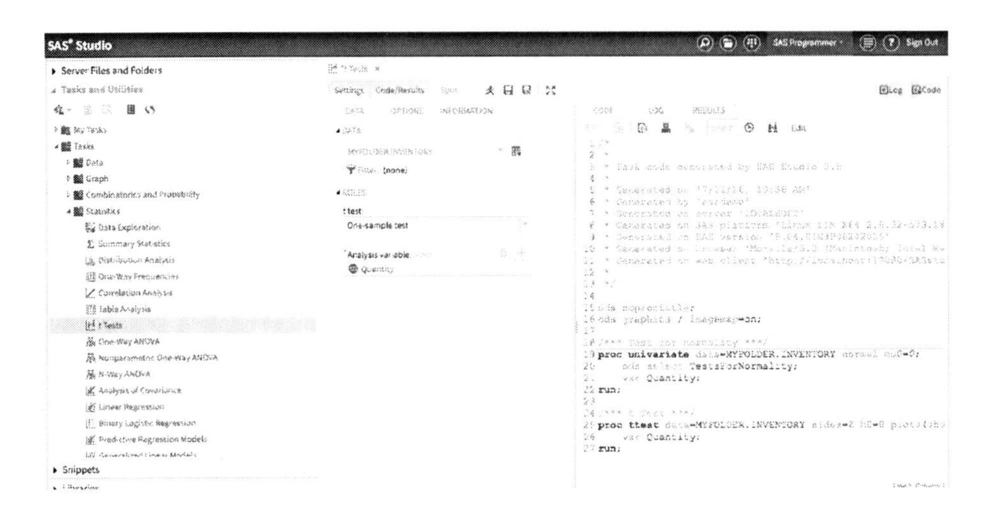

Figure 3.27 *PROC TTEST Results*

Variable: Quantity

N	Mean	Std Dev	Std Err	Minimum	Maximum
22	39.8182	10.9659	2.3379	20.0000	60.0000

Mean	95% CL Mean		Std Dev	95% CL Std Dev	
39.8182	35.7952	Infty	10.9659	8.4366	15.6710

DF	t Value	Pr > t
21	-0.08	0.5306

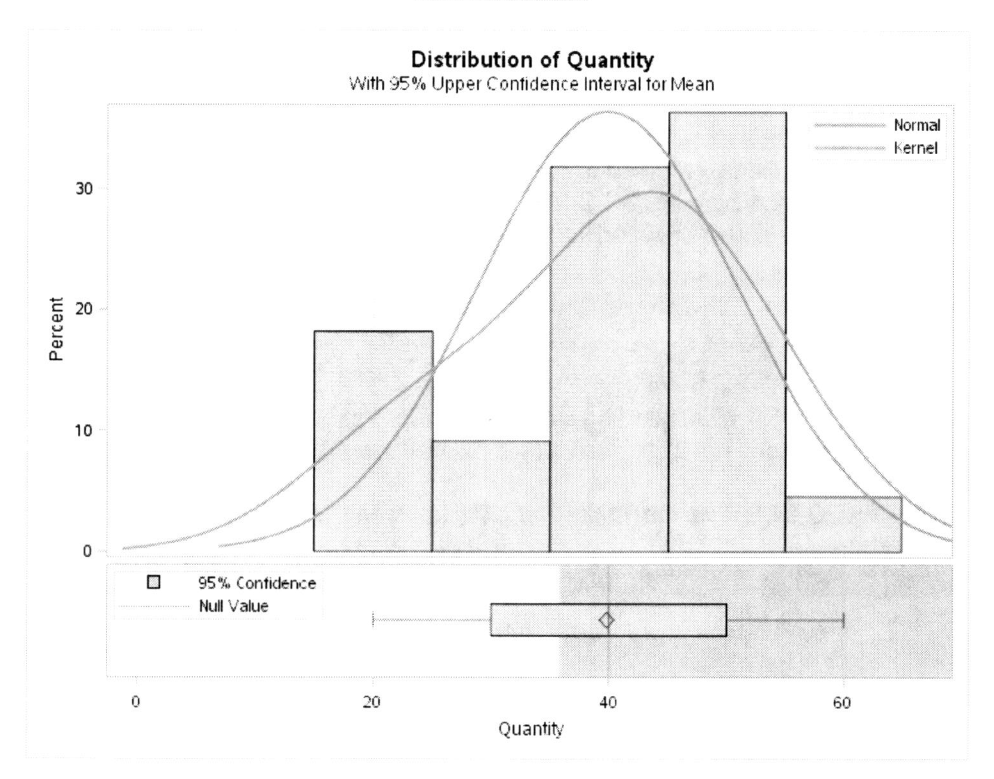

On the first line of output, you see the mean, standard deviation, standard error, minimum, and maximum for the variable Quantity. On the next line, you see the mean and the 95% confidence interval. The alpha was .05, the default. On the last line, you see the *t* value and the probability or *p*-value. With this particular *t* test, we tested to see where the mean quantity purchased was more than 40 pounds. From the results, the *p*-value was 0.53, which means the result was not significant. We cannot reject the null hypothesis that the mean quantity order was less than or equal to 40 pounds.

The histogram in Figure 3.27 on page 79 is called a summary panel. It shows the histogram for your dependent variable (Quantity) and a box plot for Quantity and its 95% confidence interval.

Further *t* tests can be used to evaluate the data. To evaluate two independent samples (groups), you use the CLASS and VAR statements. The CLASS statement determines the two groups. The VAR statement lists the response variable.

```
PROC TTEST options;
    CLASS variable;
    VAR variable;
Run;

PROC TTEST options;
    PAIRED variable1 * variable2;
Run;
```

PROC CORR

You can use PROC CORR to compute correlations between numeric variables. The correlation analysis provides a method to measure the strength of a linear relationship between two numeric variables. PROC CORR can be used to compute a Pearson product-moment correlation coefficient between variables.

```
PROC CORR data=myfolder.candy;
    var TotalFat SatFat Cholesterol Sodium Carbohydrate Sugars;
    with calories;
run;
```

You can run PROC CORR by expanding **Statistics** under **Tasks and Utilities**, selecting **Correlation Analysis**, and not writing any code. The results are the same.

Figure 3.28 *PROC CORR via the Tasks and Utilities Menu*

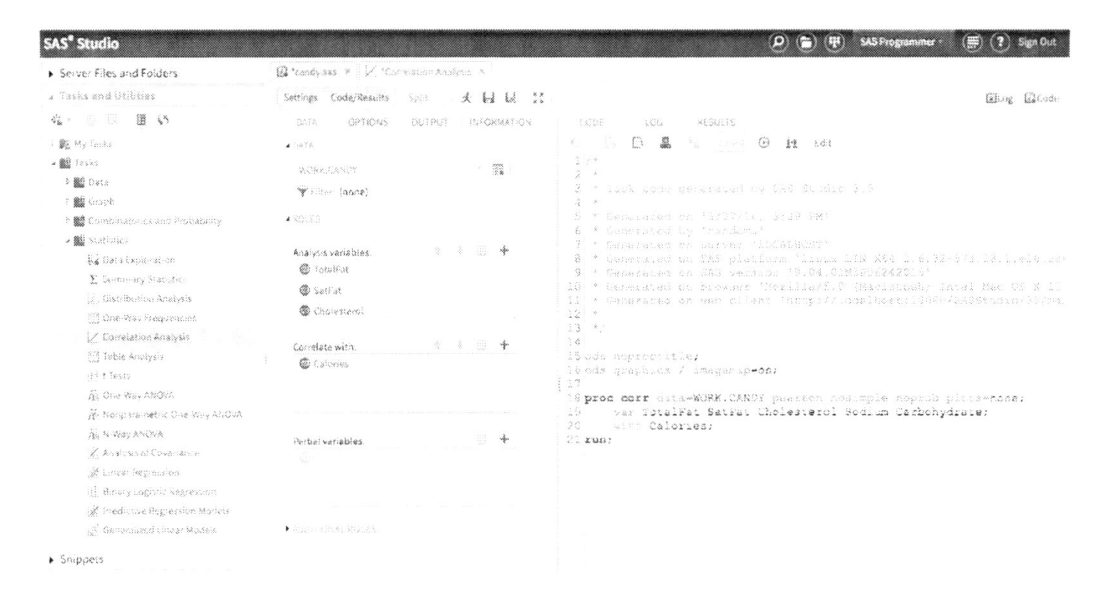

Figure 3.29 *PROC CORR Results*

The CORR Procedure

1 With Variables:	Calories
6 Variables:	TotalFat SatFat Cholesterol Sodium Carbohydrate Sugars

Simple Statistics						
Variable	N	Mean	Std Dev	Sum	Minimum	Maximum
Calories	75	233.88573	68.89958	17541	1.63000	420.00000
TotalFat	74	20.69595	57.37952	1532	0	450.00000
SatFat	74	6.45270	4.34959	477.50000	0	29.00000
Cholesterol	75	5.40000	5.40770	405.00000	0	20.00000
Sodium	75	74.33333	50.27743	5575	0	210.00000
Carbohydrate	74	32.12162	11.93585	2377	17.00000	95.00000
Sugars	75	24.26667	8.16607	1820	1.00000	48.00000

Pearson Correlation Coefficients Prob > |r| under H0: Rho=0 Number of Observations						
	TotalFat	SatFat	Cholesterol	Sodium	Carbohydrate	Sugars
Calories	-0.47075 <.0001 74	0.08968 0.4473 74	0.05487 0.6401 75	0.25160 0.0294 75	-0.05586 0.6364 74	0.22381 0.0536 75

When you specify a VAR statement and a WITH statement, you generate a Pearson product-moment correlation coefficient between every variable in the VAR list with every variable in the WITH list.

Output shows the correlation of the variable Calories with each of the other variables. In the Pearson Correlation Coefficients table, the first number in each column is the Pearson product-moment correlation coefficient. The second number is the *p*-value for the test of the null hypothesis that the population coefficient is equal to 0. The Pearson product-moment correlation coefficients are represented with the letter "r." Values of r range from -1.0 to +1.0. With positive correlation coefficients, as one variable's value increases, the other variable's value tends to increase. With negative correlation coefficients, as one variable's value increases, the other variable's value tends to decrease.

The true unknown population is represented with rho *p*. The *p*-value is a function of the observed sample results that is used for testing the statistical hypothesis. Specifically, the *p*-value is defined as the probability of obtaining a test statistic equal or more extreme than what was actually observed. Before the test is conducted, a threshold value is established, called the significance level. The significance level is usually 5% or 1%. If you choose a significance level of 5%, this level implies a 5% chance of concluding that the correlation coefficient is different from 0.

If the *p*-value is less than the significance level, then reject the null hypothesis and conclude that the correlation coefficient is significantly different from 0. If the *p*-value is greater than the significance level, then fail to reject the null hypothesis and conclude that there is not enough evidence to determine the correlation coefficient.

In Figure 3.29 on page 81, 74 observations make up each correlation. If there were missing values, and each pair of variables had a different N, the sample size would have been listed with each correlation.

Regression

Regression is a versatile analytical tool. Regression modeling can help with explaining relationships between variables. It can be used to make predictions about the response variable. The regression line approximates the relationship between X and Y. It is the line that fits the data best. Suppose you want to analyze how much total fat influences calories in a candy bar. You can run a simple linear regression model, PROC REG. The variables in PROC REG must be numeric.

```
PROC REG data=myfolder.candy;
    Model total_fat=calories;
Run;
```

You can run PROC REG by expanding **Statistics** under **Tasks and Utilities**, selecting **Linear Regression**, and not writing any code. The results are the same.

Figure 3.30 *PROC REG via the Tasks and Utilities Menu*

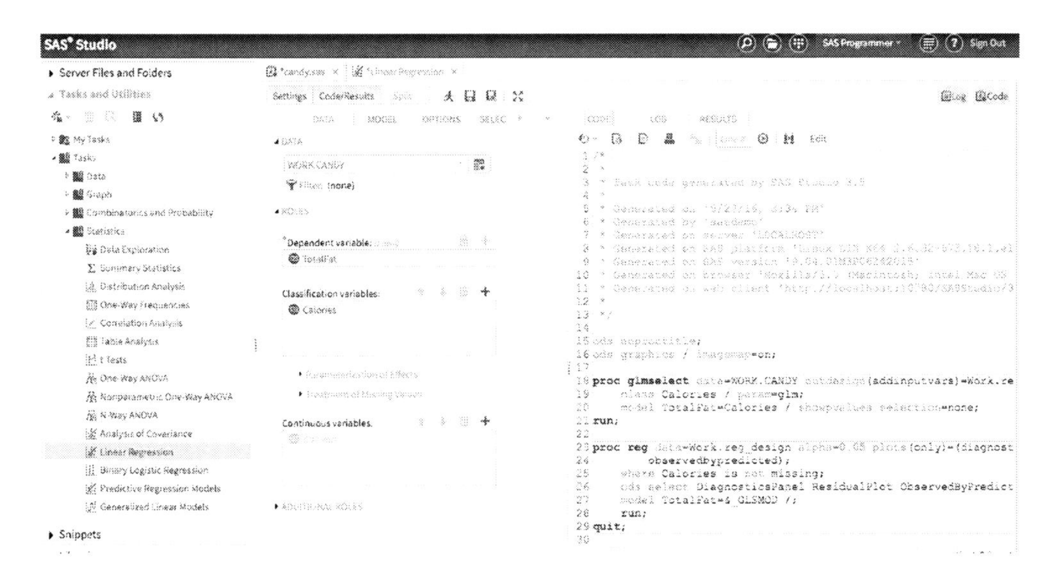

Figure 3.31 *PROC REG Results*

The REG Procedure
Model: MODEL1
Dependent Variable: TotalFat

Number of Observations Read	75
Number of Observations Used	74
Number of Observations with Missing Values	1

Analysis of Variance

Source	DF	Sum of Squares	Mean Square	F Value	Pr > F
Model	1	53262	53262	20.50	<.0001
Error	72	187084	2598.38388		
Corrected Total	73	240346			

Root MSE	50.97435	R-Square	0.2216
Dependent Mean	20.69595	Adj R-Sq	0.2108
Coeff Var	246.30111		

Parameter Estimates

Variable	DF	Parameter Estimate	Standard Error	t Value	Pr > |t|
Intercept	1	111.78959	20.97454	5.33	<.0001
Calories	1	-0.38939	0.08601	-4.53	<.0001

The first part of the output shows the number of observations read and the number of operations that were used to calculate the results. In the second part, you see the standard Analysis of Variance table, which shows the source of variance, degrees of freedom, sum of squares, mean square, *F* value, and *p*-value.

Next, you see the root mean square error, the R-square, and the adjusted R-square. The root mean square error is the square root of the mean square error. It represents variance in the system due to error. The R-square represents the proportion of variability in the dependent variable that can be explained by the regression model. The adjusted R-square adjusts for the number of predictor (independent) variables in the model, enabling you to compare models that have different numbers of predictors.

The Parameter Estimates table shows the standard error and the *t* value and *p*-value.

Figure 3.32 *Regression Fit Plot*

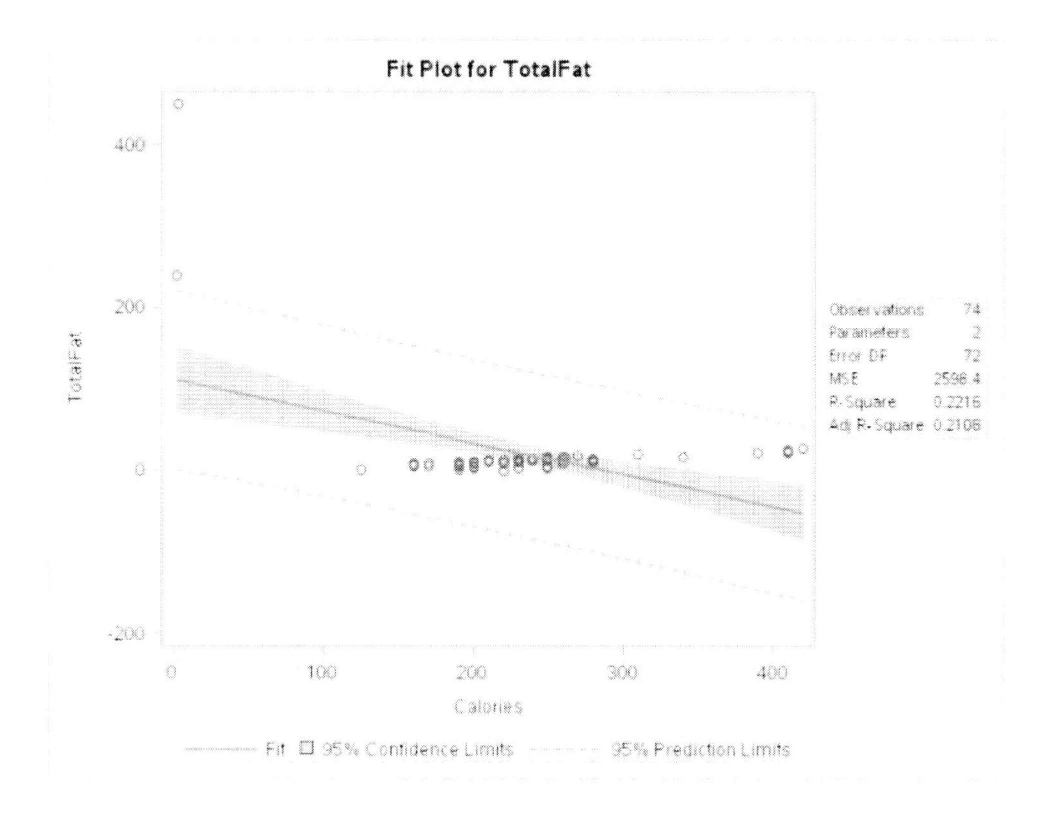

Analysis of Variance (ANOVA)

The analysis of variance (ANOVA) is a statistical test that analyzes the differences among the means of groups. In its simplest form, ANOVA provides a statistical test of whether the means of several groups are equal. PROC ANOVA has two required statements: the CLASS statement and the MODEL statement. The following sample is the general format for the code:

```
PROC ANOVA data=data set;
    CLASS variable;
    Model dependent=effects;
    Means effect/options;
Run;
```

You can run PROC ANOVA by expanding **Statistics** under **Tasks and Utilities**, selecting **One-Way ANOVA**, and not writing any code. The results are the same.

There are other PROC ANOVA options if One-Way ANOVA is not your preference.

Figure 3.33 *PROC ANOVA via the Tasks and Utilities Menu*

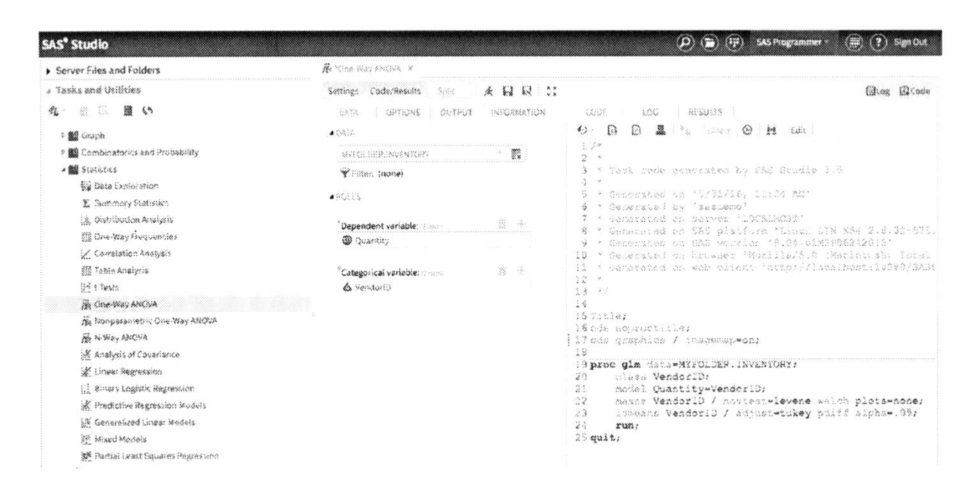

The CLASS statement must come before the MODEL statement. It defines the classification variables. For PROC ANOVA, you can use both numeric and character variables.

You can find more information about applying statistics in SAS in the book *Essential Statistics Using SAS® University Edition* by Geoff Der and Brian S. Everitt.

Quick Tips

1 *Analytics* is extracting information out of data, and *statistics* is the study of the collection, analysis, interpretation, presentation, and organization of data.

2 A graph is a visual display of data that presents frequency distributions. In a graph, the shape of the distribution can easily be seen.

3 PROC UNIVARIATE provides in-depth analysis for the distribution, whereas PROC MEANS provides a more precise summary of statistics.

4 PROC FREQ counts the frequency at which a variable appears in a data set.

5 PROC CORR computes correlations between numeric variables.

4

Real-World Case Studies

Introduction

A recipe is the guide to create a delectable dish, and that dish tells a story. Different ingredients work together to create a palatable and harmonious taste. Apple pie is an American classic.

Figure 4.1 *All American Apple Pie*

There are so many different ways to make an apple pie or a variation of apple pie. There are recipes for apple strudel, apple cobbler, apple crisp, apple squares, and even apple scruddle. Case studies are the same; there are many ways to complete a project. A case study looks in detail at a subject. But, there can be many different case studies to describe a single subject, just as there can be many different ways to make an apple pie. I like my apple pie very simple, with apples, sugar, cinnamon, some pats of butter, and a store-bought crust!

Figure 4.2 *Recipes from My Grandmother's Recipe Book*

This chapter explores a few case studies. It might give you an idea of a project that you want to implement. The case studies included are:

1. Homecoming Project
2. World Statistics Day
3. Color Flower Project

4 Food Journal Project

5 A Chef's Life and North Carolina Agriculture

6 Student Technology Use

7 New Lunch Survey Project

8 Shuttle Stops and the Magnet Program

9 The Analytics behind an NBA Name Change

10 StatSquad: Sports by the Numbers

Case Study Examples

Project Setup

These case studies are examples of some projects from my work with SAS. As you review them, think of how you could apply any relevant information to your work. All of the projects include:

1. Collecting or receiving data.
2. Accessing data with SAS.
3. Using simple analytical procedures.
4. Presenting the data.

The case studies range from simple to complex, but they all started with a question.

Case Study 1: Homecoming Project

At the school where I taught, students were complaining about Homecoming week. They were not exactly excited about Homecoming week or the school spirit days leading up to the Homecoming game, so I challenged the students to make a change. We started this project as we did other projects—we asked ourselves some questions.

- What is the question?
- How do we collect the data?
- How will we analyze the data?

The students created the following survey:

1 What class do you represent?

2 What spirit day are you looking forward to the most?

3 What is your favorite decade?

4 Of which academy[1] are you a member?

5 What is your favorite school color?

6 What would you like to have for the next Homecoming week?

7 What is your favorite sport?

8 What genre of music do you like the most?

9 Please specify your age.

10 How many hours of sleep do you get during the week?

To gather data for the survey, iPads were used during lunch hour. Students encouraged other students to take the survey by handing out M&M'S. M&M'S were significant because they are a well-known part of SAS culture. Overall, they collected data from 160 students.

They imported the data into SAS using a CSV (comma-separated values) file. They used basic summary statistics to analyze their work. Before importing the data, however, they worked together as a class to review the raw data and ensure that it was ready to be imported. This was a small exercise in cleansing the data.

```
PROC MEANS data = myfolder.homecoming; */this runs summary statistics on
                                        numeric variables*/
run;

PROC FREQ data = myfolder.homecoming; */this runs frequencies on all
                                       variables*/
run;
```

Figure 4.3 on page 92 shows the results from the survey named Favorite Spirit Day.

1 The school is a technology magnet school.

Figure 4.3 *Homecoming Results for Favorite Spirit Day*

Favorite Spirit Day				
FavSpiritDay	Frequency	Percent	Cumulative Frequency	Cumulative Percent
GQ Day	47	29.38	47	29.38
Retro Day	38	23.75	85	53.13
Spirit Day	47	29.38	132	82.50
Tacky Day	28	17.50	160	100.00

Case Study 2: World Statistics Day

To honor World Statistics[1] day in October, a project was created to see which soda students really liked best. There is a long debate over Coke versus Pepsi in the Carolinas. (It complicates matters that Pepsi is Carolina-born!) Therefore, we wanted to know which one people really prefer. A small survey was created and several classes collaborated to implement the project. The SAS class created and designed the project and survey. Participants were given a taste test and completed the survey. The results were imported into SAS for analysis, and the results were passed to the E-Commerce class to be placed on a web page created by students.

The students asked the following survey questions:

1. Please choose your age.
2. Please choose your grade.
3. Check your gender.
4. Which soda do you like best? Coke or Pepsi?
5. Which soda did you like best after taste-testing? Coke or Pepsi?

The survey was 99% completed before the taste test. Then, participants tasted the sodas and completed the rest of the survey. The survey was created using a Google form. The results were downloaded to an Excel file. The SAS class students used their coding

1 To find out more about World Statistics day, visit https://worldstatisticsday.org.

knowledge to import the data into SAS to find results. Below are a few of the procedures that they used to analyze the data:

```
Data myfolder.tastetest;
    infile '/folders/myfolders/tastetest.csv' dlm=',';
    Input age grade gender: $6. prelikesoda: $7. postlikesoda: $3.;
run;

PROC UNIVARIATE data = myfolder.tastetest;
    Var age;
run;

PROC FREQ data = stj.tastetest;
    Title "Coke vs Pepsi Taste Test";
    tables age grade gender prelikesoda postlikesoda;
    label postlikesoda="A = Pepsi B = Coke";
run;
```

Figure 4.4 *PROC FREQ Taste Test Results*

Coke vs Pepsi Taste Test

Age	Frequency	Percent	Cumulative Frequency	Cumulative Percent
14	32	11.59	32	11.59
15	58	21.01	90	32.61
16	68	24.64	158	57.25
17	101	36.59	259	93.84
18	17	6.16	276	100.00

Grade	Frequency	Percent	Cumulative Frequency	Cumulative Percent
9	42	15.22	42	15.22
10	50	18.12	92	33.33
11	72	26.09	164	59.42
12	112	40.58	276	100.00

Gender	Frequency	Percent	Cumulative Frequency	Cumulative Percent
Female	114	41.30	114	41.30
Male	162	58.70	276	100.00

PreLikeSoda	Frequency	Percent	Cumulative Frequency	Cumulative Percent
Coke	108	39.13	108	39.13
Neither	51	18.48	159	57.61
Pepsi	117	42.39	276	100.00

A = Pepsi B = Coke				
PostLikeSoda	Frequency	Percent	Cumulative Frequency	Cumulative Percent
A	126	49.03	126	49.03
B	131	50.97	257	100.00
Frequency Missing = 19				

To further showcase the results, the students created a visual graph of the PRELIKESODA and POSTLIKESODA variables. The graphs are bar charts that are easy to read and clearly show the dynamic results.

Figure 4.5 Using SAS Graphs to Display Findings

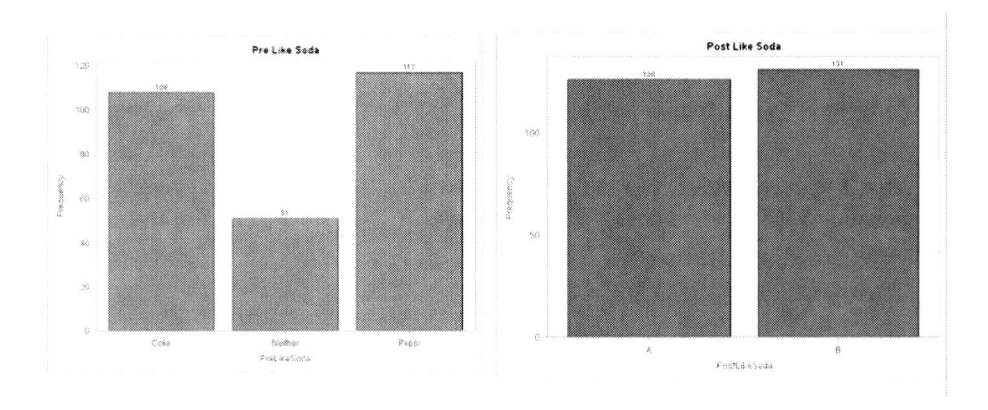

After the students in the SAS class ran simple analytics, the data was shared with the E-Commerce class to create a web page of the results.

Figure 4.6 Web Page of Taste Test Data

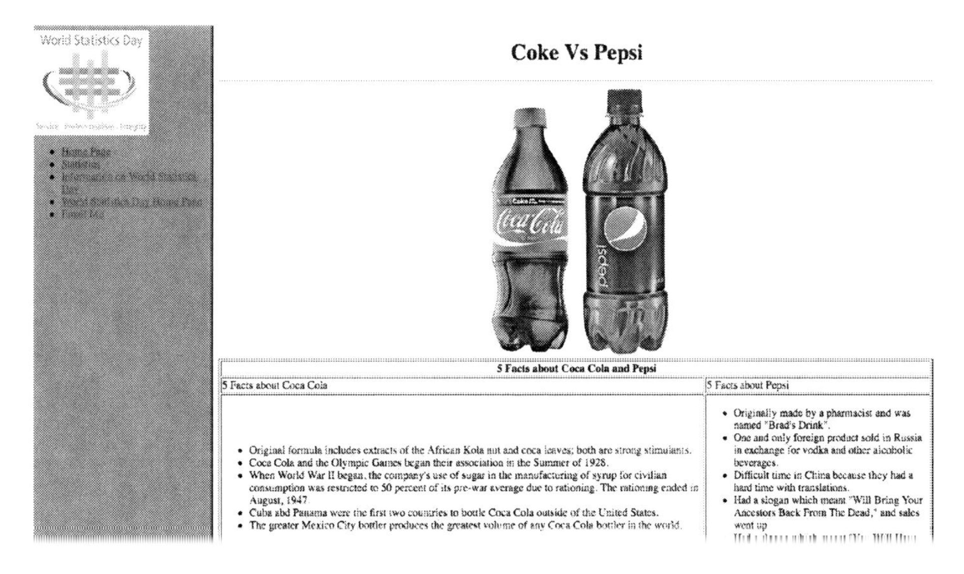

The following year, students conducted a taste test between Lay's potato chips and Utz potato chips. The project had the same setup. The SAS class students took the collected data and ran simple analytics, and the E-Commerce students created a web page to display the results.

Case Study 3: Color Flower Project

Using an easy and fun color science project by Anne Marie Helmenstine, Ph.D.[1], a SAS data analysis project was created. The science portion of the project involved students trimming the stems of flowers so that they weren't excessively long, making a slanted cut at the base of the stem under water[1], adding food coloring to a glass, and placing the flower in the glass to see whether the color changed. Students used SAS to analyze the results. Below is an example of the data collection spreadsheet that was used for the project.

Table 4.1 *Data Collection Spreadsheet Example*

Name	What is your favorite flower?	What color did you choose?	Do you think it will change color?	How long do you think it will take to change color? (record in hours)

```
PROC FREQ data = myfolder.colorflower;
Tables favcolor colorchoose willchange;
Run;
```

If you wanted to, this simple project could be enhanced to include a full scientific process in which a null hypothesis is set, and correlations, regression, or a simple *t* test are run.

1 Anne Marie is the chemistry expert for chemistry.about.com.

1 Cutting underwater prevents air bubbles from forming in the tiny tubes at base of the stem, which prevent water or color from being drawn up.

Figure 4.7 Color Flower Project

Case Study 4: Food Journal Project

The Food Journal project was executed as part of Health Sciences week. It involved observing the food intake of students for a week. They recorded what they ate each day. We ran SAS to see how the data correlated with nutrition. The goal of the project was for students to see how data is a part of their everyday lives. Students used the chart in Figure 4.8 on page 98 to record their food intake for a week along with their activities. In this project, the data was limited to the information on the form, but the form could be changed to include other content and to run different analyses.

Figure 4.8 *Food Journal Example*

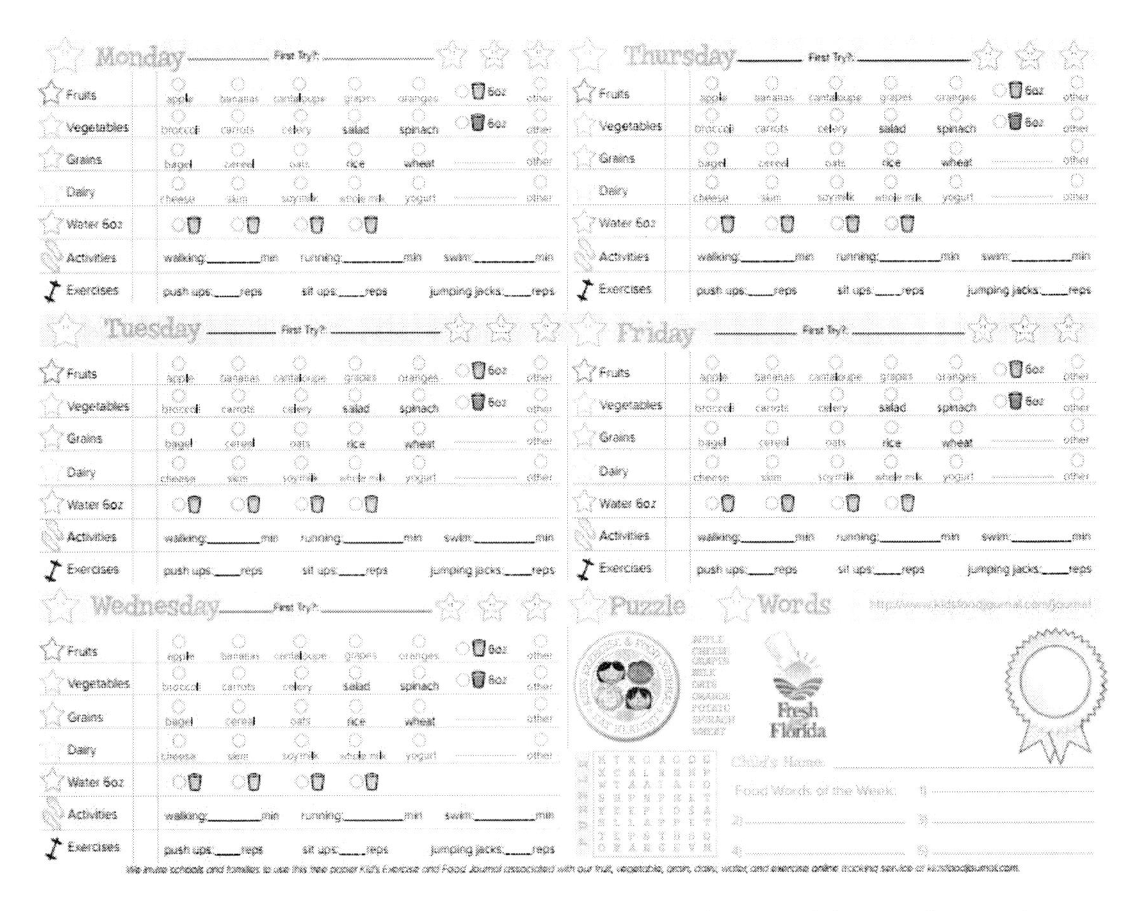

Monday ______ First Try?: ______

Fruits	apple	bananas	cantaloupe	grapes	oranges	6oz	other
Vegetables	broccoli	carrots	celery	salad	spinach	6oz	other
Grains	bagel	cereal	oats	rice	wheat		other
Dairy	cheese	skim	soymilk	whole milk	yogurt		other
Water 6oz							
Activities	walking: ______ min	running: ______ min	swim: ______ min				
Exercises	push ups: ____ reps	sit ups: ____ reps	jumping jacks: ____ reps				

Thursday ______ First Try?: ______

Fruits	apple	bananas	cantaloupe	grapes	oranges	6oz	other
Vegetables	broccoli	carrots	celery	salad	spinach	6oz	other
Grains	bagel	cereal	oats	rice	wheat		other
Dairy	cheese	skim	soymilk	whole milk	yogurt		other
Water 6oz							
Activities	walking: ______ min	running: ______ min	swim: ______ min				
Exercises	push ups: ____ reps	sit ups: ____ reps	jumping jacks: ____ reps				

Tuesday ______ First Try?: ______

Fruits	apple	bananas	cantaloupe	grapes	oranges	6oz	other
Vegetables	broccoli	carrots	celery	salad	spinach	6oz	other
Grains	bagel	cereal	oats	rice	wheat		other
Dairy	cheese	skim	soymilk	whole milk	yogurt		other
Water 6oz							
Activities	walking: ______ min	running: ______ min	swim: ______ min				
Exercises	push ups: ____ reps	sit ups: ____ reps	jumping jacks: ____ reps				

Friday ______ First Try?: ______

Fruits	apple	bananas	cantaloupe	grapes	oranges	6oz	other
Vegetables	broccoli	carrots	celery	salad	spinach	6oz	other
Grains	bagel	cereal	oats	rice	wheat		other
Dairy	cheese	skim	soymilk	whole milk	yogurt		other
Water 6oz							
Activities	walking: ______ min	running: ______ min	swim: ______ min				
Exercises	push ups: ____ reps	sit ups: ____ reps	jumping jacks: ____ reps				

Wednesday ______ First Try?: ______

Fruits	apple	bananas	cantaloupe	grapes	oranges	6oz	other
Vegetables	broccoli	carrots	celery	salad	spinach	6oz	other
Grains	bagel	cereal	oats	rice	wheat		other
Dairy	cheese	skim	soymilk	whole milk	yogurt		other
Water 6oz							
Activities	walking: ______ min	running: ______ min	swim: ______ min				
Exercises	push ups: ____ reps	sit ups: ____ reps	jumping jacks: ____ reps				

Puzzle Words

Fresh Florida

Child's Name: ______

Food Words of the Week: 1) ______ 2) ______ 3) ______ 4) ______ 5) ______

After the students collected their data for a week, they created a spreadsheet and entered their data. Figure 4.9 on page 99 shows an example of the entered data.

Figure 4.9 *Food Journal Spreadsheet*

Day	Fruits	Vegetables	Grains	Water	Activity	ActivityMins	Exericse	ExerciseReps
Monday	apple	carrots	cereal	6	Walking	30	pushup	10
Tuesday	grapes	broccoli	cereal	8	Walking	20	pushup	15
Wednesday	grapes	other	cereal	12	Walking	20	pushup	20
Thursday	grapes	salad	cereal	18	Walking	20	pushup	25
Friday	apple	salad	cereal	18	Walking	30	pushup	25
Saturday	apple	salad	cereal	24	swim	60	none	
Sunday	apple	salad	cereal	12	none		pushup	20

Descriptive statistical procedures were run, including PROC PRINT, PROC FREQ, and PROC MEANS. A correlation analysis was performed and conditional statements were used. Students merged their data sets.

```
PROC MEANS data = myfolder.foodjournal;
    Title "Summary Statistics for Food Journal";

run;
TITLE; */this is a null Title statement that negates the Title statement
from above from the remaining PROC statements*/

PROC CORR data = myfolder.foodjournal;
    var activitymins exercisereps;
    with water;
run;
```

Figure 4.10 *Summary Statistics for Food Journal Project*

Summary Statistics for Food Journal

Variable	Label	N	Mean	Std Dev	Minimum	Maximum
Water	Water	7	14.0000000	6.3245553	6.0000000	24.0000000
ActivityMins	ActivityMins	6	30.0000000	15.4919334	20.0000000	60.0000000
ExerciseReps	ExerciseReps	6	19.1666667	5.8452260	10.0000000	25.0000000

Figure 4.11 *PROC CORR Results for Food Journal Project*

The CORR Procedure

1 With Variables:	Water
2 Variables:	ActivityMins ExerciseReps

Simple Statistics

Variable	N	Mean	Std Dev	Sum	Minimum	Maximum	Label
Water	7	14.00000	6.32456	98.00000	6.00000	24.00000	Water
ActivityMins	6	30.00000	15.49193	180.00000	20.00000	60.00000	ActivityMins
ExerciseReps	6	19.16667	5.84523	115.00000	10.00000	25.00000	ExerciseReps

Pearson Correlation Coefficients
Prob > |r| under H0: Rho=0
Number of Observations

	ActivityMins	ExerciseReps
Water Water	0.63960 0.1712 6	0.97598 0.0009 6

You can run these procedures by expanding the **Tasks and Utilities** menu as shown in Chapter 3. SAS Studio generates the code, and the results are the same as if you wrote the code yourself.

Students used a conditional statement to implement a point system for the food that they ate during the week. A conditional statement is based on the IF-THEN/ELSE statement. In the following code, two variables are created to calculate points for fruit and vegetables. Points are assigned based on nutritional value. For the project, students had to research the nutritional value of each of the foods from the food journal.

```
data work.foodjournalpoints;
    set myfolder.foodjournal;

    if fruits in ("apple" "bananas" "oranges") then
        FruitPoint=3;
    else if fruits in ("grapes" "cantaloupe") then
        FruitPoint=2;
    else
        FruitPoint=0;

    if vegetables in ("broccoli" "carrots") then
        VegPoint=3;
    else if vegetables in ("celery" "salad") then
        VegPoint=2;
```

```
    else if vegetables in ("spinach") then
        VegPoint=4;
    else
        VegPoint=0;
run;

PROC PRINT data = work.foodjournalpoints;
    title 'Food Journal Analysis';
    sum fruitpoint vegpoint;
run;
```

Figure 4.12 *Conditional Statements Results*

Food Journal Analysis

Obs	Day	Fruits	Vegetables	Grains	Water	Activity	ActivityMins	Exericse	ExerciseReps	FruitPoint	VegPoint
1	Monday	apple	carrots	cereal	6	Walking	30	pushup	10	3	3
2	Tuesday	grapes	broccoli	cereal	8	Walking	20	pushup	15	2	3
3	Wednesday	grapes	other	cereal	12	Walking	20	pushup	20	2	0
4	Thursday	grapes	salad	cereal	18	Walking	20	pushup	25	2	2
5	Friday	apple	salad	cereal	18	Walking	30	pushup	25	3	2
6	Saturday	apple	salad	cereal	24	swim	60	none	.	3	2
7	Sunday	apple	salad	cereal	12	none	.	pushup	20	3	2
										18	14

Students were tasked with merging their data with other students' data to compare. The first step was to load other students' data into their own folder and library. Then, they merged data sets together to create one new data set.

```
PROC SORT data myfolder.foodjournal;
    By day;
Run;
Data myfolder.foodjournalmerge;
    Merge myfolder.foodjournal myfolder.foodjournal1;
Run;
```

After the data sets were merged, students worked as a team to analyze the data and prepare a presentation.

Case Study 5: A Chef's Life and North Carolina Agriculture

North Carolina is still an agriculture state. Over half of the state is rural, and farming is an important economic force in the state. Because the concept of going local is very popular, this case study looked at the agricultural aspects of North Carolina. We watched the show, *A Chef's Life*, which has already been mentioned in this book, to see how one chef is taking the ingredients of the South and creating cuisine. The website www.achefslifeseries.com describes the show as "a Peabody and Emmy award-winning docu-series that plunges audiences into the kitchen of a high-end restaurant located in the low country of eastern North Carolina. It follows the trials and travails of Chef Vivian Howard and her husband, Ben Knight, and their farm-to-table restaurant, Chef & the Farmer, exploring both traditional and modern applications of quintessential Southern ingredients."

I chose two episodes to watch to give context to our goal. We began our research of North Carolina agriculture by looking at the North Carolina Agriculture Overview website at http://www.ncagr.gov/stats/general/overview.htm. On the site are links to agriculture statistics. This site provides insight on crops grown in and sold from North Carolina. We explored the United States Department of Agriculture National Agricultural Statistics Service at https://www.nass.usda.gov/. On this site, we reviewed national statistics on agriculture. Students exported data from one or both of the sites. They ran analytics on the data that they collected from these public resources. The data was saved in the Work library. Remember, if a library name is not specified in the DATA statement, the default library is Work.

```
Data  cornimport;
    infile '/folders/myfolders/corncsc.csv' dlm=',' dsd;
    input year Period: $4. commodity: $4. AcresPlanted  AcreHarvested YieldBUperAcre;
run;

PROC SORT data = cornimport out = cornsorted;
    by year;
    run;

PROC SGSCATTER data = cornsorted;
      compare x=(year AcresPlanted) y=(YieldBUperAcre AcreHarvested) /
      markerattrs=graphdata1(symbol=circlefilled);
run;
```

This code can be manually written or selected from the **Tasks and Utilities** menu. Figure 4.13 on page 103 and Figure 4.14 on page 104 show both scatter plots.

Figure 4.13 *Scatter Plot From Manually Written Code*

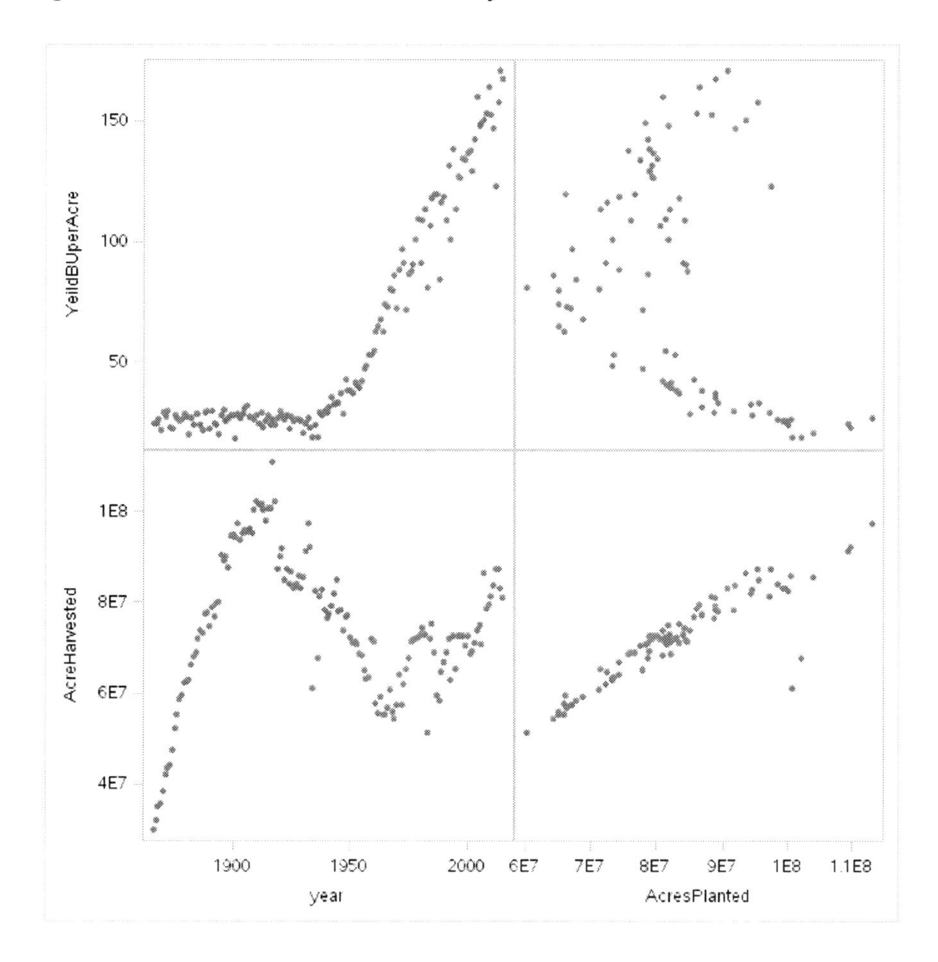

Figure 4.14 *Scatter Plot via the Tasks and Utilities Menu*

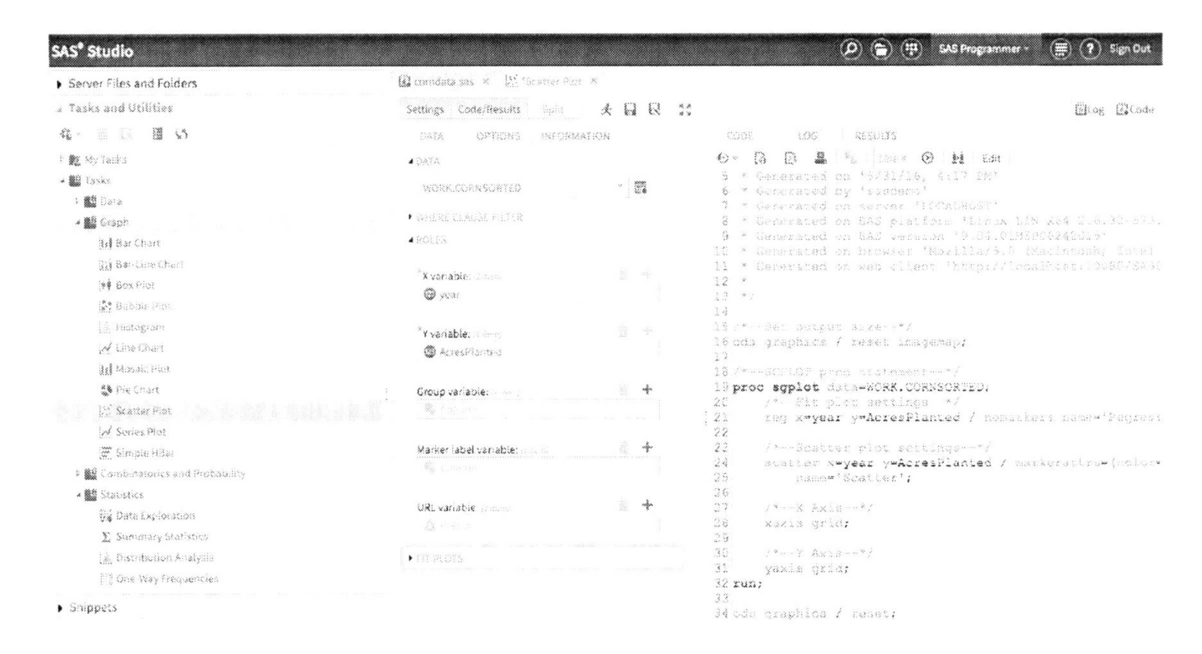

Case Study 6: Student Technology Use

To better understand how students use technology and to determine whether more educational content could be communicated via technology, a small survey was conducted. Students were asked the following questions:

1. What type of cellphone do you use?
2. What type of tablet do you use?
3. Do you prefer a laptop or desktop? Which operating system do you use?
4. What is your primary social network?
5. How do you communicate with your friends?
6. How do you communicate with your family?
7. How do you communicate with classmates?
8. How would you like Berry[1] to communicate with you?
9. What is your preferred coursework communication?

---

1 The Phillip O. Berry Academy of Technology.

10 What is your primary technology device?

Based on the use of summary statistics, the following results were found:

- 45% of the students used an Android; 41% of the students used an iPhone.
- 57% of the students used a laptop with a Windows operating system.
- 63% of the students used Instagram as their main social media site.
- 74% of the students communicated via text, but 33% said they would like the high school to communicate via email.
- 40% of the students said they would like to receive coursework via hard copy.

```
PROC PRINT data = myfolder.studenttechuse;
Run;

PROC MEANS data = myfolder.studenttechuse;
    var age grade;
run;

PROC FREQ data = myfolder.studenttechuse;
    tables CellPhone Tablet ComputerUs SocialNetworks CommunicateFriend
        CommunicateFamily CommunicateClassmate CommunicateBerry
         PreferCoursework TechnologyUseMost;
run;
```

Figure 4.15 *PROC FREQ Student Technology Use Results*

The FREQ Procedure

CellPhone				
CellPhone	Frequency	Percent	Cumulative Frequency	Cumulative Percent
Option 6	7	2.94	7	2.94
phone- Android Phone	106	44.54	113	47.48
phone- Blackberry	8	3.36	121	50.84
phone- Simple cell phone	9	3.78	130	54.62
phone- Windows Phone	11	4.62	141	59.24
phone- iPhone	97	40.76	238	100.00

Tablet				
Tablet	Frequency	Percent	Cumulative Frequency	Cumulative Percent
None	48	20.17	48	20.17
ereader- kindle	3	1.26	51	21.43
ereader- nook	7	2.94	58	24.37
tablet- android	44	18.49	102	42.86
tablet- ipad	107	44.96	209	87.82
tablet- kindle fire	14	5.88	223	93.70
tablet- windows	15	6.30	238	100.00

ComputerUse				
ComputerUse	Frequency	Percent	Cumulative Frequency	Cumulative Percent
desktop - windows	51	21.43	51	21.43
desktop- mac	9	3.78	60	25.21
laptop- linux	5	2.10	65	27.31
laptop- mac	38	15.97	103	43.28
laptop- windows	135	56.72	238	100.00

***Figure 4.16** PROC FREQ Student Technology Use Results (continued)*

SocialNetworks				
SocialNetworks	Frequency	Percent	Cumulative Frequency	Cumulative Percent
Facebook	28	11.76	28	11.76
Foursquare	2	0.84	30	12.61
Google+	5	2.10	35	14.71
Instagram	149	62.61	184	77.31
None	12	5.04	196	82.35
Tumblr	8	3.36	204	85.71
Twitter	34	14.29	238	100.00

CommunicateFriend				
CommunicateFriend	Frequency	Percent	Cumulative Frequency	Cumulative Percent
Email	7	2.94	7	2.94
Hard copy	9	3.78	16	6.72
Phone call	40	16.81	56	23.53
Social media site message	7	2.94	63	26.47
Text message	175	73.53	238	100.00

CommunicateFamily				
CommunicateFamily	Frequency	Percent	Cumulative Frequency	Cumulative Percent
Email	11	4.62	11	4.62
Hard copy	5	2.10	16	6.72
Phone call	142	59.66	158	66.39
Social media site message	2	0.84	160	67.23
Text message	78	32.77	238	100.00

CommunicateClassmate				
CommunicateClassmate	Frequency	Percent	Cumulative Frequency	Cumulative Percent
Email	13	5.46	13	5.46
Hard copy	8	3.36	21	8.82
Phone call	19	7.98	40	16.81
Social media site message	11	4.62	51	21.43
Text message	187	78.57	238	100.00

Figure 4.17 *PROC FREQ Student Technology Use Results (continued)*

CommunicateBerry				
CommunicateBerry	Frequency	Percent	Cumulative Frequency	Cumulative Percent
Email	80	33.61	80	33.61
Hard copy	38	15.97	118	49.58
Phone call	47	19.75	165	69.33
Social media site message	10	4.20	175	73.53
Text message	63	26.47	238	100.00

PreferCoursework				
PreferCoursework	Frequency	Percent	Cumulative Frequency	Cumulative Percent
depends	81	34.03	81	34.03
digital	61	25.63	142	59.66
hard copy	96	40.34	238	100.00

TechnologyUseMost				
TechnologyUseMost	Frequency	Percent	Cumulative Frequency	Cumulative Percent
Computer	15	6.30	15	6.30
Gaming Device	14	5.88	29	12.18
None	7	2.94	36	15.13
Phone	184	77.31	220	92.44
Tablet	18	7.56	238	100.00

For other graph choices, in SAS Studio, expand the **Tasks and Utilities** menu, and select **Graph**. You are offered a variety of visual charts to choose from to see the relationships between these variables.

Figure 4.18 Box Plot for Age and Social Media

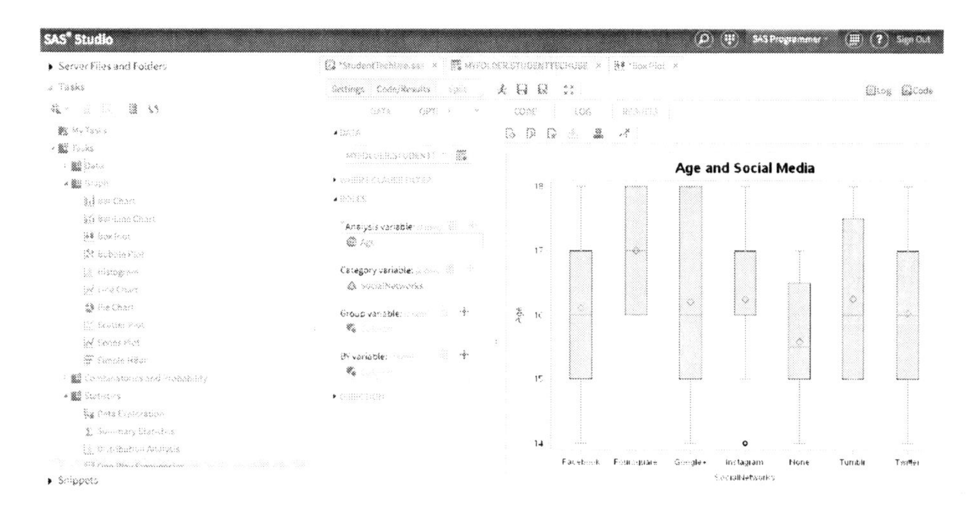

In Figure 4.18 on page 109, a box plot dives deeper into the breakdown of age and social media. There is a large age group that uses Google+, although a much more concentrated group uses Instagram. These results tell a story about the use of social media at the high school and how technology can be used to enhance students' educational experiences.

Case Study 7: New Lunch Survey Project

The New Lunch Survey project was one that the students decided to take on themselves. They wanted to collect student opinions on what students thought about the new lunch schedule. During the 2012 school year, a new lunch schedule was put into effect, moving students from four 20-minute lunch periods to two 40-minute lunch periods.

Students created a 10-question survey about the new lunch schedule. After data was collected, they used SAS to analyze it.

The survey consisted of 10 opinion-based questions:

1. What grade are you in?
2. Do you like the new lunch schedule?
3. What lunch do you have?
4. Do you find it hard to get from the lunch area to class?
5. Where do you eat lunch the most?
6. Is 40 minutes too long for lunch?
7. Did it take you a long time to get to lunch?
8. How do you like spending your lunch?

9 Would you prefer only one lunch?

10 Would you like music to be played during lunch?

The data collected from the survey questions was imported into SAS using a CSV file.

The following figure shows the project instructions and the outline given to students:

Figure 4.19 *Project Instructions*

SAS Programming
New Lunch Survey Project

Part 1

I. Open the Excel File , input the LunchSurvey data
II. Save as CSV file
III. Create a SAS data set from your CSV file (what is the delimiter?)
IV. Create a report of your data. Include
  a. Title1: Phillip O. Berry
  b. Title4: New Lunch Survey
  c. Create permanent descriptive labels (where will these be placed)
  d. Format your data appropriately for readability

V. Run a Proc Print
VI. Run a Proc Sort - You choose the variables based on what you want to see sorted (i.e. by grade, lunch, lunchperiod, etc)
VII. Run a Proc Print for your sorted data set- use a where statement or only display certain variables
VIII. Run a Proc Means for grade variable display mean, min, max
  a. What is the average grade that took the survey? ______________
IX. Run a Proc Freq for at least FOUR variables
X. Chart at least two of the variables

Part 2

I. Create a new data set from your original LunchSurvey data set
II. Use Keep/Drop to keep or drop variables
III. Proc Print the new data sets you create
IV. Run two other procedures on the new data set
V. Create a chart

Part 3

I. Create a presentation from the data analyzed
II. Print code and staple to the back of these instructions

The general statistics from the results showed:

- 250 students took the survey.
- Of the 250, the average grade was 10th grade.

- 49% of students eat and socialize at lunch, but 48% of students said they would do all of the above, which is eat, go to the library, and make up work.
- 87% said that 40 minutes is not too long for lunch.

```
PROC MEANS data = myfolder.lunchsurvey;
run;

PROC FREQ data = myfolder.lunchsurvey;
    label grade="Student Grade"
        getlunch="Did it take you a long time to get to lunch?"
        music="Would you like music to be played during lunch?";
run;
```

Figure 4.20 *Lunch Survey PROC FREQ Results*

Student Grade				
Grade	Frequency	Percent	Cumulative Frequency	Cumulative Percent
9	56	22.58	56	22.58
10	67	27.02	123	49.60
11	70	28.23	193	77.82
12	55	22.18	248	100.00

Do You Like the New Lunch?				
Lunch	Frequency	Percent	Cumulative Frequency	Cumulative Percent
no	32	13.06	32	13.06
yes	213	86.94	245	100.00
Frequency Missing = 3				

Which Lunch Do You Have?				
LunchPeriod	Frequency	Percent	Cumulative Frequency	Cumulative Percent
1st	116	47.35	116	47.35
2nd	129	52.65	245	100.00
Frequency Missing = 3				

LunchtoClass	Frequency	Percent	Cumulative Frequency	Cumulative Percent
no	220	88.71	220	88.71
yes	28	11.29	248	100.00

Where Do You Eat The Most?				
Eat	Frequency	Percent	Cumulative Frequency	Cumulative Percent
cafeteria	93	38.43	93	38.43
commons area	67	27.69	160	66.12
outside	82	33.88	242	100.00
Frequency Missing = 6				

Figure 4.21 *Lunch Survey PROC FREQ Results (continued)*

Is 40 Minutes Too Long For Lunch?				
LunchTime	Frequency	Percent	Cumulative Frequency	Cumulative Percent
no	211	87.55	211	87.55
yes	30	12.45	241	100.00
Frequency Missing = 7				

Did it take you a long time to get to lunch?				
GetLunch	Frequency	Percent	Cumulative Frequency	Cumulative Percent
no	203	82.86	203	82.86
yes	42	17.14	245	100.00
Frequency Missing = 3				

Where Would You Spend Your Lunch?				
SpendLunch	Frequency	Percent	Cumulative Frequency	Cumulative Percent
All of the Above	120	48.39	120	48.39
Eating/Socializing	123	49.60	243	97.98
Library	3	1.21	246	99.19
Making Up Work	2	0.81	248	100.00

Would You Like To Have Only One Lunch?				
OneLunch	Frequency	Percent	Cumulative Frequency	Cumulative Percent
no	172	69.64	172	69.64
yes	75	30.36	247	100.00
Frequency Missing = 1				

Would you like music to be played during lunch?				
Music	Frequency	Percent	Cumulative Frequency	Cumulative Percent
no	17	6.88	17	6.88
yes	230	93.12	247	100.00
Frequency Missing = 1				

This case study was a starting point for students to learn the art of collecting data and seeking answers. This starting point could be expanded to calculate sample size and include more in-depth variables.

Case Study 8: Shuttle Stops and the Magnet Program

Magnet schools are a part of many public school systems. They offer valuable opportunities for students interested in specialized programs. Phillip O. Berry Academy of Technology, a magnet school in Charlotte, NC, is a technology and STEM magnet.

In 2010, the district decided to implement shuttle stops to bus students to magnet schools. As a result, students no longer had buses come to their neighborhoods to take them to their magnet schools. Instead, they had to provide their own transportation to the shuttle stops. Berry was affected by this change, and the SAS class was given an opportunity to find out its true impact. The students created a 12-question survey for Berry students to collect their current transportation. In addition, they completed the design process to obtain an accurate sample. They surveyed 606 students and imported the data into SAS to analyze.

Figure 4.22 *Newsletter about the Project*

Phillip O. Berry

May 7, 2010

Satellite Busing:

Where Does Phillip O. Berry Stand?

SAS Class Conducts Transportation Survey

The SAS programming class was given the opportunity by the POB School Leadership Team to conduct a transportation survey. The leadership team was interested to find out how our students were currently getting to school and from school and if satellite busing would have a tremendous impact on the school.

The SAS class then developed a 12 question survey to ask students about their current transportation to school. The survey was then posted on the school website.

The major portion of data collection was done on April 28th during the two lunch periods. Laptop computers were set up in the cafeteria for students to participate in the survey. Data collection was successful with [illegible] students taking the survey.

Students then returned to class to analyze the data using SAS.

Please see the back side for the overall findings from our Survey.

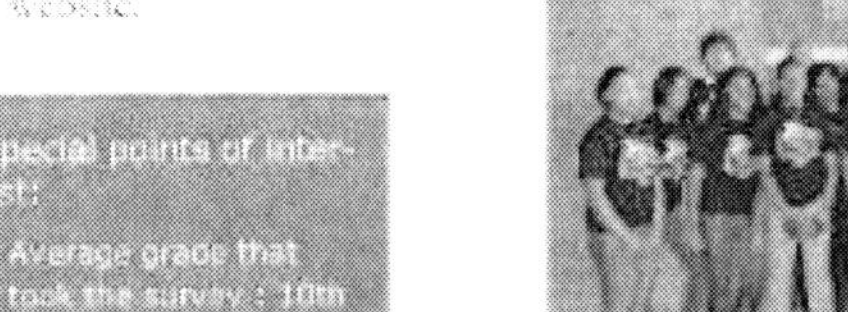

Special points of interest:

- Average grade that took the survey : 10th grade
- Zip codes that represent the study body:
    - 28208
    - 28216
    - 28214
    - 28269
    - 28215

Figure 4.23 *Survey Results*

Survey Results

1. **How do you currently get to school?**
   School Bus: 85%
   Parents: 8%
   Drive Yourself: 5%
2. **How do you currently get home from school?**
   School Bus: 85%
   Parents: 7%
   Drive Yourself: 5%
3. **Could you get to a satellite location if it was within 3 miles of your house?**
   No: 75%
4. **Could you get home from a satellite location if it was within 3 miles of your house?**
   No: 80%
5. **Could you get to school without a school provided bus?**
   No: 61%
6. **Could you get home from school without a school provided bus?**
   No: 64%
7. **Would you have to go back to your home school if you could not get to POB?**
   Yes: 60%
8. **Would you feel safe at a satellite bus location?**
   Depend on Location: 42%
   No: 50%
   Yes: 41%

"I was surprised to see such involvement from the student body. It seems that students are concerned about getting to school."

-Kierra Staggers, SAS class 11th grade

The students presented the results to the Berry administration team and to the Charlotte School Board. You can see their presentation at http://bit.ly/1VtPjSY.

Figure 4.24 *Students before the School Board Presentation*

After the survey was conducted, the Berry administration team asked the SAS class to create a survey for parents. They wanted to find out the parents' perceptions of the shuttle stops as well.

Case Study 9: The Analytics behind an NBA Name Change

In the spring of 2013, a high school SAS Programming class had a discussion about the Charlotte Bobcats. The Bobcats were the professional basketball team in Charlotte, North Carolina. There had been some grassroots efforts on social media to change the name of the Bobcats to a more meaningful name. For Charlotte, the Hornets were the professional basketball team from 1988 to 2002. The then-owner moved the Hornets to New Orleans, and the Charlotte Bobcats were formed in Charlotte. Over the years, the fan base for the Bobcats was never really strong because there just wasn't a connection. So, the SAS class wondered aloud, "If the Bobcats changed its team name, would the dwindling fan base return?"

They created a survey that consisted of 10 questions that covered topics like, "Do you like the name 'Bobcats'?" "Do you attend basketball games?" "Do you buy merchandise?"

Specifically, the survey questions were:

1 How old are you?

2 Please check your gender.

3 Do you like basketball?

4 Do you attend Bobcats games?

5 Do you like the name "Bobcats" for the city of Charlotte's pro basketball team?

6 If you could rename the Bobcats, out of these five options, which would you select?

7 Do you buy professional basketball merchandise?

Within a one-hour class period, students from the class surveyed 981 out of 1,733 students at Phillip O. Berry Academy of Technology. Using Base SAS 9.3, after importing the data from Excel, they wrote SAS programs to format, analyze, and report on the data. The results confirmed the class' hypotheses that a name change would be positive for the team.

Students followed the instructions in Figure 4.25 on page 120 to guide their process of analyzing the data in SAS.

Figure 4.25 *Project Instructions*

SAS Programming
Bobcats Project

Part 1

I. Open the Excel File , input the Bobcats data
II. Create a SAS data set from your CSV file (what is the delimiter?)
III. Create a report of your data. Include
   a. Title1: Bobcats Game
   b. Title4: Player Stats
   c. Create permanent descriptive labels (where will these be placed)
   d. Format your data appropriately for readability

IV. Run a Proc Print that sums all the numeric variables
V. Run a Proc Sort
VI. Run a Proc Print from your sorted data set and use a where statement or only display certain variables
VII. Run a Proc Means for three of the numeric variables and only display mean, min, max
VIII. Run a Proc Freq for at least two variables
IX. Run a Proc Report choosing four variables for the columns
   a. Define the columns
   b. Order by one variable
   c. Include a break
   d. Include a headline and headskip

Part 2

I. Create a new data set from Bobcats
II. Use Keep/Drop to keep or drop variables
III. Run a Proc Means
IV. Run a Proc Print

Part 3

I. Create a Presentation from the data analyzed
II. Print code and staple to the back of these instructions

Below is an example of the code one of the students wrote. Students were asked to include PROC REPORT, which is a procedure that enables a programmer to create a more customized report.

```
Data myfolder.Bobcats;
    infile '/folders/myfolders/Bobcats.csv' dlm=',';
    input Age: $7. Gender: $6. LikeBasketball: $3. AttendGames: $9. LikeName: $3.
        Change: $45. Buy: $3.;
run;
PROC REPORT data = myfolder.Bobcats headline headskip;
    column Age Gender LikeBasketball AttendGames LikeName Change Buy;
    define Age / order width=7;
    define Gender / width=6;
    define LikeBasketball / width=14;
    define AttendGames / width=11;
    define LikeName / width=8;
    define Change / width=45;
    define Buy / width=3;
    title1 Bobcats Project ;
    title4 Project Results;
run;
PROC MEANS data = myfolder.Bobcats;
run;
PROC SORT data = myfolder.Bobcats  out = work.Bobcatssort;
    by Gender;
run;
PROC PRINT data = work.Bobcatssort;
    var Gender AttendGames LikeName;
run;
PROC FREQ data = work.Bobcatssort;
    tables LikeName * Buy / crosslist;
run;
PROC FREQ data = wor.Bobcatssort;
    tables Gender * Buy / crosslist;
run;
```

The students presented their results to several Charlotte Bobcats representatives, who stated they had spent millions of dollars for what the students had done simply using SAS. And, after all of the data was collected and presented, the Charlotte Bobcats did change its name to the Charlotte Hornets. The Charlotte Hornets were the original NBA team in Charlotte until the team moved to New Orleans.

The following year, the students conducted a follow-up survey to see whether opinions had changed after the name changed back to the Charlotte Hornets. They collected data in the same manner and worked in SAS to analyze the results.

The follow-up survey questions were:

1 How old are you?

2. Please check your gender.
3. Did you know that the Charlotte Bobcats name has been changed to the Charlotte Hornets?
4. Are you happy with the name change?
5. Are you more likely to attend games now that the name has changed?
6. Now that the name has changed, are you more willing to buy merchandise?
7. How much are you willing to pay for merchandise?
8. Which merchandise are you more likely to buy?
9. Do you think the Hornets should have new colors?
10. Have the Bobcats done enough to engage your interest with the name change?

Figure 4.26 *Students Lauren Cook and Jeremiah Kelly Presenting the Project at the NC Science Teachers Association PDI Conference*

The students ran their data using the SAS Code Editor. However, all of the data could have been run using the **Tasks and Utilities** menu in SAS University Edition.

The project was presented at the North Carolina Science Teachers Association PDI Conference and SAS Global Forum 2015. The SAS Global Forum 2015 poster is available at http://support.sas.com/resources/papers/proceedings15/3278-2015.pdf. This project began as a grassroots idea of the students. The strategy could easily be applied to any question that might be circling around on social media.

Case Study 10: StatSquad: Sports by the Numbers

The StatSquad STEM Program is an after-school athletic-based program that provides learning opportunities and hands-on experience for high school students in the areas of information technology, digital media, and data analytics. The program was designed by National Amateur Sports. The program shows how all curricula can be infused with STEM. Through the program, students are exposed to the data collection of sports statistics. Students at one high school decided to take the program a step further. They collected statistics from three different sport teams: football, basketball, and baseball. They entered the data into SAS for analysis. Students presented the results to coaches, community members, and administration. The results showed that these teams were being recruited to do sporting events more than other sport teams. The coaches found these results especially valuable.

Figure 4.27 *Students Jeryn Lindsay, Tierra Faulkner, and Joseph Williams Presenting Data*

One team that students worked closely with over the course of the season was the tennis team. Below is some sample code from working with the tennis team:

```
Title "Tennis";
data sasstats.tennis;
   infile"tennis.csv" dlm= ',' MISSOVER firstobs=2
input Date: mmddyy10.
        Opponent: $32.
        first_name: $10.
        last_name: $10.
        location: $10.
        condition: $10;
format date mmddyy10.;
run;
PROC PRINT data = sasstats.tennis noobs;
run;
```

The process of collecting data, processing the data, and working through the analysis exposed students to true integration of analytics and sports. Students learned more about the sport teams and how to use data to strengthen the players and the teams.

More Ideas

Here are a few more ideas for projects that you can do in your own community:

- Collect data on higher education possibilities in your area (for example, four-year universities, two-year universities, or trade schools).
- Use the Nielsen ratings website to extract data on top popular culture. For more information, visit http://www.nielsen.com/us/en.html.
- Use data from a local fundraiser.
- Pull data from your local professional sports team.
- Collect data on fashion trends.
- Collect data on holiday food to find out your local population's favorite culinary traditions.

Figure 4.28 *A Slice of Apple Pie*

These are just a few ideas. Data storytelling, much like making a recipe, is the art of giving meaning to the individual pieces that make up a data set. Data is everywhere, and the ability to craft data into knowledge can be as easy as apple pie. Simplicity is the key; you have to start small and then grow.

SAS is a great analytical tool because once you learn the syntax, the data part is easy. You can always stay relevant with your work because the syntax of SAS, DATA statements, and PROC statements stays consistent. You just add different data. You can use any data that is current and relevant to craft a meaningful project. You can create your own recipe for success; just ask a question and then tell the story!

Quick Tips

1 Start simple.

2 Tell a story!

3 One project can form into two!

4 Use the **Tasks and Utilities** menu to deepen your analysis.

5 Make it relevant!

Index

A

B

C

H

I

K

L

M

N

O

P

Q

R

S

T

U

V

W

CPSIA information can be obtained at www.ICGtesting.com
Printed in the USA
LVOW09s0606190816

500697LV00005B/14/P